TRANSFORMING U.S. ENERGY INNOVATION

One of the greatest challenges facing human civilization is the provision of secure, affordable energy without causing catastrophic environmental damage. As the world's largest economy, and as a world leader in energy technologies, the United States is a particularly important case. In the light of increased competition from other countries (particularly China), growing concerns about the local and global environmental impacts of the energy system, an ever-present interest in energy security, and the realization that technological innovation takes place in a complex ecosystem involving a wide range of domestic and international actors, this volume provides a comprehensive and analytical assessment of the role that the U.S. government could play in energy technology innovation. It will be invaluable for policy makers in energy innovation and for researchers studying energy innovation, future energy technologies, climate-change mitigation, and innovation management. It will also act as a supplementary textbook for courses on energy and innovation.

LAURA DIAZ ANADON is Assistant Professor of Public Policy at the John F. Kennedy School of Government, Harvard University, and Associate Director of the Science Technology and Public Policy Program at the Belfer Center for Science and International Affairs. Her research focuses on energy- and environment-oriented technological progress and on the role of government policy. She also studies the coupling between energy and water infrastructure systems and their implications for decision making. Her articles have been published in a variety of journals, including *Environmental Science & Technology*, *Energy Economics*, *Research Policy*, and *Issues in Science and Technology*. Before Harvard she published in chemical engineering and nuclear magnetic resonance journals, worked on research at DuPont and Bayer Pharmaceuticals and at Johnson Matthey Catalysts, and worked as a financial consultant for banks on credit risk models for financing technology projects. She was a Speaker on the 2013 U.S. National Academy of Engineering Frontiers of Engineering Symposium, and on the advisory board of the Accelerating Energy Innovation Project at the International Energy Agency and has been a consultant for various international organizations and advised government officials around the world on policy for energy innovation.

MATTHEW BUNN is Professor of Practice at the John F. Kennedy School of Government, Harvard University. His research interests include policies for energy innovation, nuclear theft and terrorism, nuclear proliferation and measures to control it, and the future of nuclear energy and its fuel cycle. Before Harvard, Bunn served as an adviser to the White House Office of Science and Technology Policy, as a study director at the National Academy of Sciences, and as editor of *Arms Control Today*. He is the author or coauthor of more than twenty books and major technical reports and more than a hundred articles in publications ranging from *Science* to the *Washington Post*. He is an elected Fellow of the American Association for the Advancement of Science, a recipient of the Joseph A. Burton Forum Award from the American Physical Society, and the recipient of the Hans A. Bethe Award from the Federation of American Scientists for science in service to a more secure world. He serves on the Nuclear Energy Advisory Committee of the U.S. Department of Energy.

VENKATESH (VENKY) NARAYANAMURTI is the Benjamin Peirce Professor of Technology and Public Policy and Professor of Physics at Harvard University. He has previously served as director of the Solid State Electronics Research Laboratory at Bell Labs; Vice President for Research and Exploratory Technology at Sandia National Laboratories; Dean of Engineering at the University of California, Santa Barbara; and founding dean of the School of Engineering and Applied Sciences at Harvard University. He has been elected to memberships in the National Academy of Engineering, the American Academy of Arts and Sciences, and the Royal Swedish Academy of Engineering Sciences. He has served on numerous advisory boards for the federal government, research universities, national laboratories, and industry. These have included chair of the Inertial Confinement Fusion Advisory Committee of the U.S. Department of Energy; chair of the Committee of Visitors of the Division of Materials Research, National Science Foundation; member of the President's Council for the University of California Managed National Laboratories; member of the Governing Board of Brookhaven National Laboratory; and the Brains Trust of ARPA-E of the U.S. Department of Energy.

TRANSFORMING U.S. ENERGY INNOVATION

Edited by

LAURA DIAZ ANADON
John F. Kennedy School of Government, Harvard University

MATTHEW BUNN
John F. Kennedy School of Government, Harvard University

VENKATESH NARAYANAMURTI
John F. Kennedy School of Government, School of Engineering and Applied Sciences, and Department of Physics, Harvard University

CAMBRIDGE
UNIVERSITY PRESS

32 Avenue of the Americas, New York, NY 10013-2473, USA

Cambridge University Press is part of the University of Cambridge.

It furthers the University's mission by disseminating knowledge in the pursuit of education, learning, and research at the highest international levels of excellence.

www.cambridge.org
Information on this title: www.cambridge.org/9781107043718

First published 2014

Printed in the United States of America

A catalog record for this publication is available from the British Library.

Library of Congress Cataloging in Publication Data
Anadon, Laura Diaz, 1981–
Transforming U.S. energy innovation / Laura Diaz Anadon, John F. Kennedy School of Government, Harvard University, Matthew Bunn, John F. Kennedy School of Government, Harvard University, Venkatesh Narayanamurti, John F. Kennedy School of Government, School of Engineering and Applied Sciences and Department of Physics, Harvard University
pages cm
Includes bibliographical references.
ISBN 978-1-107-04371-8 (hardback)
1. Energy development – United States. 2. Renewable energy sources – United States. 3. Energy industries – United States. 4. Research, Industrial – United States. 5. Energy policy – United States. I. Title.
HD9502.U52A47 2014
333.790973 – dc23 2014001502

ISBN 978-1-107-04371-8 Hardback

Contents

1

The Need to Transform U.S. Energy Innovation

MATTHEW BUNN, LAURA DIAZ ANADON, AND
VENKATESH NARAYANAMURTI

1.1. The Need for an Energy Technology Revolution

How will countries provide the reliable, affordable energy needed to fuel a growing world economy and lift billions of people out of poverty without causing catastrophic climate change and other environmental disasters? Answering this question is perhaps the greatest challenge human civilization faces in the 21st century. It cannot be done without a revolution in the technologies of both energy production and use. And for that revolution to arrive in time will require a dramatic acceleration in the pace at which new or improved energy technologies are invented, demonstrated, and adopted in the marketplace.

This book outlines policies the U.S. federal government in particular can adopt to foster such an energy technology revolution. These approaches will help meet the myriad energy challenges of the 21st century – and help ensure that the United States plays a leading role in doing so, capturing its share of the multi-trillion-dollar energy technology markets of the coming decades.

This book seeks to answer four basic questions. First, how much should the U.S. government spend on energy research, development, and demonstration (RD&D), and how should that money be allocated? Second, how can the U.S. government best work with the private sector and encourage it to invest its own funds in researching, developing, demonstrating, and deploying new energy technologies? Third, how can the U.S. government manage its energy RD&D investments and the institutions through which they flow to get the most beneficial energy innovation per dollar invested? Fourth, how can the U.S. government best manage competition and cooperation with other countries in energy technology innovation?

There are no simple answers to any of these questions. But for each of them, we offer new data on the current situation and new analyses that help elucidate the best paths forward. Adopting new approaches in these four areas could transform the picture. With an expanded and effectively targeted federal investment in energy RD&D; effective policies to encourage private sector innovation and deployment; strengthened approaches to managing energy innovation institutions; and strong international cooperation, the United States could radically step up the pace at which it invents and deploys improved energy technologies.

This book draws on a 2011 report, *Transforming U.S. Energy Innovation* (Anadon et al., 2011). In the brief time since then, the U.S. and global energy picture has continued to change, with events ranging from continued expansion of shale gas and oil to conflict with China over trade in solar panels to deep difficulties in Japan over providing energy in the aftermath of the Fukushima nuclear accident. In this book, we have updated our arguments to reflect these events, and added new data, analyses, and case studies. Ours is by no means the first study to call for steps to accelerate U.S. energy technology innovation, including expanded federal investments in energy RD&D (see, e.g., NAS, 2009; Weiss & Bonvillian, 2009; American Energy Innovation Council, 2010, 2011; American Enterprise Institute, Brookings Institution, and Breakthrough Institute, 2010; PCAST, 2010; DOE, 2011b; Henderson & Newell, 2011; Lester & Hart, 2011; Gruebler et al., 2012). But our study provides new insights, through:

- Written surveys of more than 100 experts to explore expert views on how much government investments in RD&D could improve cost and performance of some 25 technologies in seven areas of energy technology[1]
- Extensive economic modeling to explore how such improvements in cost and performance might affect outcomes such as the cost of the U.S. energy system, U.S. carbon emissions, or U.S. oil import dependence
- A broad survey of private-sector energy innovation in the United States, outlining what firms are involved and some of the key drivers of their energy innovation decisions
- An in-depth analysis of DOE's funding for energy RD&D in the private sector
- Detailed case studies of key innovation institutions, to glean lessons on how to increase the effectiveness of DOE's national laboratories and other energy innovation institutions in developing new energy technologies and getting them adopted in the market
- New data on emerging economies' investments in energy technology innovation, and on how international cooperation in energy RD&D is evolving

We integrate these elements in a systems-level approach to energy technology innovation, offering specific recommendations for action to accelerate such innovation in the United States.

[1] Bionergy (12 experts, 5 reviewers), utility scale storage (25 experts, 7 reviewers), fossil energy (12 experts, 3 reviewers), nuclear (30 experts, 2 reviewers), vehicles (9 experts, 1 reviewer), and photovoltaic power (11 experts, 1 reviewer).

Some of the needed steps are already being taken. The Obama administration, the Bush administration before it, and the U.S. Congress have all recognized the pressing need for major improvements in energy technologies, though they have taken different and sometimes conflicting approaches to the problem. As we describe in Chapter 2, billions of dollars in federal investments in energy innovation were a major element of the stimulus bill signed into law in 2009. The new Advanced Research Projects Agency – Energy (ARPA-E) – intended to offer a new focus on high-risk, high-payoff ideas – was authorized in 2007 and funded in 2009. New energy innovation "hubs" have been established to push progress in particular technology areas, from computer simulation of nuclear reactors to making liquid fuels from solar energy, and "Energy Frontier Research Centers" have been created to focus on particular challenges at earlier stages of the innovation chain (DOE, 2013a, 2013b). Loan guarantee programs (established during the Bush administration but invigorated by the Obama administration) have supported construction of new nuclear reactors and new firms in areas such as battery technology and solar power (to some considerable controversy, particularly after the collapse of the solar firm Solyndra). New efficiency standards and incentives, tax credits, and portfolio standards have helped drive deployment of efficiency measures and low-carbon energy technologies, encouraging further private investment in innovation in these areas. Secretary of Energy Steve Chu and his successor Ernie Moniz (chair of the 2010 President's Council of Advisors on Science and Technology [PCAST] study) have both focused intensely on accelerating energy technology innovation. Figure 1.1 shows the main government programs and institutions – old and new – with a primary mandate to spur energy innovation. Whether and how all of these initiatives will gain support and work together remains uncertain.

But deep divisions over political, economic, and technical strategies have prevented the U.S. government from adopting and implementing either a coherent long-term energy strategy in general or an integrated energy innovation strategy in particular (PCAST, 2010; Deutch, 2011). With the current focus on budget cutting in the United States, bipartisan support for energy R&D spending has broken down, with committees in the U.S. Congress proposing substantial reductions in funding. Philosophically, many in the U.S. Congress believe that the government should fund early stage research and then get out of the way and let private markets do the rest – and many are not convinced that human-caused climate change is occurring or is a serious problem (e.g., Inhofe, 2012). These disagreements have led to substantial political fights over issues ranging from loan guarantees to carbon constraints to efficiency standards.

It is important to understand that the incumbent fossil fuel technologies are deeply entrenched, with trillions of dollars in investment already made, decades of cost-reducing innovation already accomplished, a wide range of institutions and infrastructure put in place to support them (both public and private), and ownership

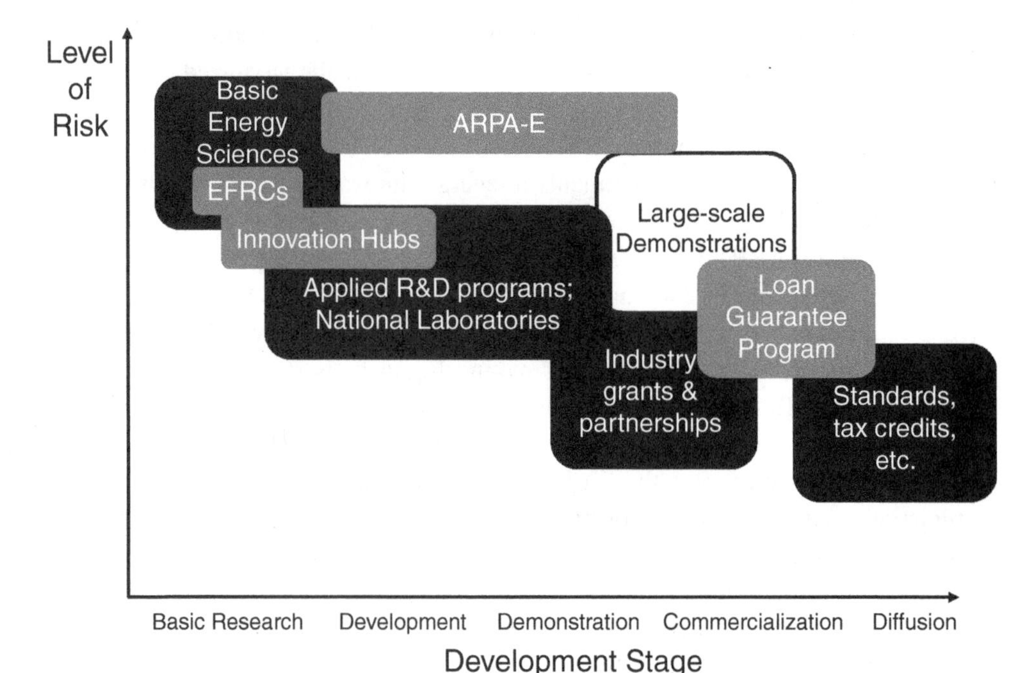

Figure 1.1. Schematic of U.S. government institutions and policies with a primary mandate to accelerate energy innovation. Black boxes denote programs that existed before 2005. The other institutions (with the exception of the "large scale demonstrations," which is an area we have identified as a gap) are more recent: the EFRCs, the Hubs, and ARPA-E were funded in 2009, and the loan guarantee program was instituted in 2005, although the first loan guarantee was issued in 2009. (Adapted from various presentations from the U.S. Department of Energy.)

by some of the world's largest and most powerful firms (Unruh, 2000). Other technologies will not find it easy to compete – and the growth and expansion of other energy technologies is likely to be accompanied by jarring economic shifts and political disagreements. Technology innovation inevitably involves creative destruction, with both winners and losers – and the potential losers are often highly effective in blocking innovations that threaten them (Acemoglu & Robinson, 2012).

In short, although the United States has strengthened its energy technology innovation efforts in recent years, a great deal remains to be done. Filling the remaining gap is likely to require rebuilding bipartisan consensus on the need for action and the roles the federal government should play (though there is a great deal other actors can do in the meantime). We do not have any answers on how this bipartisan consensus can be rebuilt. But we hope that the analysis and evidence presented in this book can help make the case for action and lay out steps the U.S. government could take to accelerate the pace of progress in energy innovation and ensure that these efforts provide the maximum return on the investments made.

1.2. Interlinked Challenges

Today, the United States is enjoying an energy bonanza. Advances in the technologies of hydraulic fracturing ("fracking") and horizontal drilling have enabled huge new production of natural gas and oil, leading to natural gas prices only a fraction of those in Asia or even Europe, and predictions that U.S. oil production may surpass Saudi Arabia's within a decade (IEA, 2012). At the same time, as a result of fuel switching from coal to natural gas and (to a lesser extent) improved efficiency and expanded use of renewables, U.S. carbon emissions are declining (IEA, 2012). Many Americans, if told there was an energy crisis, would respond: "Crisis? What crisis?"

Unfortunately, these developments have, in effect, only lengthened the fuse on the deep and difficult challenges energy will still pose for the United States and globally in the decades to come. In some respects, this progress on fossil fuels has made a more fundamental transformation of energy production and use more difficult: cheap natural gas is making all other energy technologies, from nuclear to renewable to efficiency, less economically attractive than they were before. Yet a more fundamental transformation of energy production and use is still urgently needed, for multiple reasons. The energy problem is simultaneously a security, an economic, a poverty, and an environmental challenge.

1.2.1. A Security Challenge

A wide range of current and former senior national security officials have made it clear that both the U.S. and global addictions to fossil fuels pose severe security challenges (CNA, 2009; Daniel, 2010). Although the increases in U.S. and Canadian production are reducing U.S. reliance on imported oil, prices in global oil markets are tightly linked (as a substantial infrastructure is in place to transport oil from one market to another). Even if the United States did not use a drop of Middle Eastern oil, the U.S. economy would remain vulnerable to sudden price increases in the event of a major disruption in oil supply in the Middle East or elsewhere.

Despite the growth in North American production, it is still the case that much of the world's reserves of oil recoverable at reasonable prices are concentrated in some of the most politically volatile regions of the world. Indeed, the International Energy Agency (IEA) projects that world reliance on the Organization of Petroleum Exporting Countries (OPEC) will increase again through the 2030s (IEA, 2012, pp. 81–124). While the mutual dependence of suppliers and consumers makes it difficult to wield the "oil weapon," chaos in oil-producing countries can scare markets and reduce production, sending prices soaring. High energy prices fill the coffers of some of the world's most hostile governments.

Moreover, military forces around the world are deeply dependent on fossil fuels that compromise operations and raise security concerns – as the casualties from fuel supply trucks in Iraq and the terrorist attacks on fuel shipments in Pakistan to provide fuel to NATO troops in Afghanistan make clear (Defense Science Board, 2008; Simeone, 2009; BBC, 2010).

With most population growth taking place in those regions with limited energy access or low energy use, an energy market that continued to be vulnerable to sharp shifts in oil prices could also lead to civil unrest in some locations and could threaten national and international security. More importantly, it is becoming increasingly clear that the natural disasters, refugee flows, and intensification of poverty that will result in some regions if climate change is not addressed will pose risks to international peace and security (CNA, 2007).

In addition, a global expansion of nuclear power without improvements in technology and strengthened institutional controls could increase the danger of nuclear proliferation and terrorism (as well as the risks of nuclear accidents).

1.2.2. An Economic Challenge

Until recently, the United States sent $700 million dollars a day overseas just to purchase petroleum, an amount representing roughly half the U.S. trade deficit (U.S. Census Bureau, 2010). Volatile oil and gas prices can disrupt investment planning, drive businesses into bankruptcy, and push families into sudden poverty.

At the same time, without an accelerated approach to energy technology innovation, the United States is in danger of losing out on the huge energy technology markets of the coming decades. The IEA projects $37 trillion in global investment in energy supply – not counting technologies of energy end-use – by 2035 (IEA, 2012, p. 49); these markets will drive huge numbers of jobs and large transfers of wealth. Where once the United States was the undisputed world leader in new energy technologies, other countries are now making large investments and rapid progress: China is now the world's largest producer of both solar cells and wind turbines; China and South Korea are able to build nuclear power plants at substantially lower costs than U.S. or European firms are able to do; and China, Europe, Japan, and South Korea, among others, are investing heavily in developing new energy technologies, as described in Chapter 5.

1.2.3. A Poverty Challenge

Today, an estimated 1.3 billion people have no access to electricity. Some 2.6 billion have no access to clean cooking facilities. Providing affordable, reliable, and clean energy for the world's poor is a moral and practical imperative. Lack of energy supplies perpetuates poverty and exacerbates the suffering poverty entails. In particular, the indoor smoke from fires in huts in the developing world is a far larger source of death and disease than all the outdoor pollution in the world's most polluted cities combined (Lim et al., 2012; Rehfuess, Corvalan, & Neira, 2006). Both improved energy technologies and improved institutional approaches to energy supply and use are needed to provide energy for the world's poorest.

1.2.4. An Environmental Challenge

Finally, in a world that remains dependent on fossil fuels for more than 80% of its energy supply, the most profound energy challenge of all is finding ways to provide the energy the world needs without causing catastrophic climate change or imposing other immense environmental burdens. From oil spills in the ocean to air pollution choking many cities around the world to global climate disruption, production and use of energy causes many of the worst environmental problems the world faces, at local, regional, and global scales.

The climate challenge is particularly pressing. The evidence is increasingly clear: the climate is warming, glaciers are melting, the seas are rising, and it appears that extreme weather events – floods, droughts, and heat waves among them – are becoming more frequent and more damaging in some regions (IPCC, 2012). Ecosystems that provide crucial services are changing, and pests and diseases are spreading to new areas as the climate warms. Climate change is already reducing crop yields in some regions, and projections suggest the combination of precipitation changes and heat stress could greatly increase the danger of widespread hunger and malnutrition (World Bank, 2012, 2013; Wheeler & von Braun, 2013; and Easterling et al., 2007). The impacts of climate change are expected to disproportionately affect the world's poor, who are least prepared to cope with them.

Global emissions of greenhouse gases set records in 2011 and 2012 and are projected to continue to grow steadily as energy use continues to expand and continues to rely primarily on fossil fuels (IEA, 2013). With a growing world population and growing economies, the IEA projects that global primary energy consumption may increase to 1.5 to 3 times the current demand by 2050 (IEA, 2010a). Despite projected growth in renewable energy supplies, the IEA projects that unless policies change, use of natural gas and coal will grow more than use of any other energy sources by 2035 (IEA, 2012, p. 51). In that case, CO_2 emissions from the energy sector would increase from the record 31.6 billion metric tons in 2012 to more than 37 billion metric tons by 2035, putting the world on a course more likely to lead to warming of 4°C or more in this century than to meet the goal of holding the temperature increase to 2°C above pre-industrial temperatures. As the president of the World Bank summarized the issue in 2012, such high levels of warming "are devastating: the inundation of coastal cities; increasing risks for food production potentially leading to higher malnutrition rates; many dry regions becoming dryer, wet regions wetter; unprecedented heat waves in many regions, especially in the tropics; substantially exacerbated water scarcity in many regions; increased frequency of high-intensity tropical cyclones; and irreversible loss of biodiversity, including coral reef systems" (World Bank, 2012, p. ix). Moreover, at such levels, the probability of abrupt, severe climate disasters greatly increases. The Bank report warned that these levels of warming "must be avoided." Yet they cannot be avoided without a radical transformation of world energy supply – and that transformation is likely to require dramatic improvements in the cost and performance of low-carbon energy technologies.

The hour is already late: the IEA reports that the last opportunity to meet a 2°C target is slipping away, given the greenhouse gases the energy infrastructure already built will emit during its lifetime (IEA, 2013). The costs of meeting that target or any other will increase the longer the world delays taking substantial climate action, as more expensive high-carbon energy infrastructure gets built every year. Moreover, the greenhouse gases already in the atmosphere will cause continued warming for decades to come, even if further emissions stopped tomorrow. Nevertheless, action that begins soon and continues to ramp up throughout the 21st century can still reduce the risks of the most catastrophic projected effects of climate change.

The other environmental impacts of the current energy technologies are also very serious. China's government has officially estimated that environmental damage – most of it from energy use – cost China $230 billion in 2010, 3.5% of its total gross domestic product (GDP; Wong, 2013). Other countries' economies are also suffering substantial damage. The latest estimates suggest that fine particulates in the air kill some 3.1 million people every year, accounting for some 3.1% of global disability-adjusted life years (DALYs) lost annually (Lim et al., 2012). Indoor smoke from cooking fires in the developing world causes even more disease and death (because the concentrations to which people are exposed are so high), estimated to be in the range of 3.5 million deaths per year and 4.5% of global DALYs (Lim et al., 2012).

1.2.5. Meeting the Challenges

These huge, interlinked challenges must be met, for affordable and reliable energy supplies are the lifeblood of the economy of the United States and of the world. The world's ever-growing reliance on burning fossil fuels simply cannot be sustained – the economic, security, and environmental costs will all prove to be unacceptably high. Indeed, it may simply not be possible to meet growing projected demands for oil and gas at an acceptable cost. While there is debate over when "peak oil" will occur, and production now appears likely to grow for some time, there is little debate that, at some point in the decades to come, oil production will stop growing and eventually decline, even as energy demand continues to grow. Price spikes, supply disruptions, and political tensions over scarce supplies are likely to become increasingly common. The need to develop alternatives to fossil fuels that can be deployed at a massive scale is real and urgent.

As the IEA has put it: "Current global trends in energy supply and use are unsustainable – environmentally, economically, socially.... It is not an exaggeration to claim that the future of human prosperity depends on how successfully we tackle the... central energy challenges facing us.... What is needed is nothing short of an energy technology revolution" (IEA, 2009a). To meet the challenges, both major innovations in energy technology and new international policies and institutions will be needed. Broad international action and cooperation will be essential: no separate solution implemented within a single country, or even region, will suffice.

The scale of what must be done to meet these challenges is staggering, as the global energy system (the infrastructure of exploration facilities, transmission lines, pipelines, power plants, buildings, vehicles, etc.) is huge. As noted earlier, the IEA estimates that simply maintaining energy supply under already announced policies will require $37 trillion in energy supply investments by 2035 (approximately $1.6 trillion a year); trillions more would be needed to put the energy system on a sustainable track – though much of that additional investment would be recouped in lower fuel costs and reduced environmental damages.

This book includes estimates of the extent to which energy RD&D investments and the resulting improvements in technology can address the interlinked challenges of security, the economy, poverty, and the environment. We deal with the uncertainty in the future outcomes of incremental and radical innovation by representing the costs of meeting a particular demand (gallons of liquid fuel, electric power, and automobile) as a range of possible outcomes.

What is clear from all of the studies conducted over the past 20 years is that today's pace of energy innovation is simply not enough to meet this century's energy challenges. The world's energy infrastructure represents an investment of tens of trillions of dollars, much of it in systems that last for decades. It is like a supertanker that takes miles to change direction, no matter how energetically one turns the wheel. Even the most attractive energy technologies of the past, which offered dramatic new services – from oil to electricity – took decades of growth before they represented a substantial fraction of total energy use. Over the past 60 years, countries around the world have poured tens of billions of dollars into research and development and subsidies for nuclear energy – yet nuclear energy provides only about 6% of primary energy worldwide. Renewable sources, such as solar and wind, have been under development for decades, yet they provide an even smaller portion of the world's energy needs. A dramatic improvement in the pace of energy technology innovation and diffusion is essential. Inventing new energy technologies is not enough – these technologies must be widely deployed if they are to make a difference in meeting security, economic, and environmental energy challenges.

1.3. Expanded RD&D Is Essential – But Will Not Be Enough

Some recent reports have suggested that the world can research its way out of its energy predicament – that with enough R&D dollars, new technologies will be invented that will be so attractive that little further government action will be needed to help them sweep the old technologies aside in the energy marketplace (American Enterprise Institute, Brookings Institution, and Breakthrough Institute, 2010).[2] Our work makes clear that this view is deeply wrong, for two key reasons.

First, the challenges are so great that there will be a need to deploy *both* technologies that can compete on their own with entrenched, incumbent technologies without further government action and technologies that cannot yet overcome the market barriers to widespread deployment without some form of demand-side policies. At any given stage of technological development, there will be some technologies that can compete with fossil fuels already and others that would be competitive if externalities (such as the health effects of fine particulates or the global impacts of carbon) were internalized in the price, or that face other market barriers that require government support to overcome. The world is likely to need large-scale deployment of technologies in both categories to fuel a growing global economy while avoiding climate catastrophes. In the modeling described in Chapter 2, we found that even a dramatic expansion of federal funding for energy RD&D, and the most optimistic expert predictions of the results of that innovation investment, would simply not be enough to achieve the dramatic reductions in U.S. carbon emissions (or the dramatic reductions in U.S. oil use) likely to be needed by 2050 unless this expanded RD&D is coupled with demand-side policies that have the effect of putting a substantial price on carbon emissions. (Indeed, a broader suite of demand-side policies will ultimately be needed, given the different market barriers to deployment of low-carbon technologies that exist in different sectors.) In fact, entire classes of technology that are likely to be crucial to achieving deep reductions in emissions – such as capturing and sequestering carbon from the burning of fossil fuels – are unlikely to ever be economically competitive if the costs to society of dumping carbon into the atmosphere are not taken into account.

Second, in the absence of substantial prices on carbon and other externalities, innovation itself will be slowed, because the private sector will have little incentive to invest in developing low-carbon and low-pollution technologies. The survey of private sector energy innovation described in Chapter 4 makes it clear that expected prices and their effect on potential profits are the dominant factors affecting private sector energy innovation investment decisions. In July 2011, American Electric Power announced that it would abandon its plans to carry out a large-scale demonstration of technology to capture and sequester carbon from an existing coal plant because, in the absence of a carbon price, the concept was simply not economically viable (AEP, 2011; *The New York Times*, July 13, 2011). As this book makes clear, to accelerate the full cycle of

[2] In addition to large RD&D investments, this three-institution report calls for several other steps designed to provide limited support for initial deployments of new energy technologies, including initial military procurements and a reform of deployment subsidies designed to focus on driving down costs. The report holds open the possibility of only a very small carbon price, designed to fund RD&D rather than to give the private sector strong incentives to deploy low-carbon technologies.

innovation from invention to deployment will require substantial government support on both the "supply side" and the "demand side" of energy technologies (Stavins, 2010).

Of course, in recommending major new federal investments in energy innovation, we are conscious of the U.S. federal deficit and the intense political disputes over government spending that continue in the United States. But just as a private firm borrows to make investments that are critical to its future profitability, the U.S. government must not neglect critically needed, long-term investments in the rush to balance the budget. As the PCAST report suggested, it may make sense to achieve a substantial portion of the needed investment through new approaches and revenue streams outside the usual appropriations process – as was done in the past when the Federal Energy Regulatory Commission approved a small surcharge on interstate transportation of natural gas that was used to finance gas RD&D (PCAST, 2010, p. 16). But one way or another, a central theme of this book is that the United States must make a larger investment in new energy technologies if it is to meet its energy challenges.

1.4. The Need for a Portfolio of Improved Energy Technologies

A major part of these challenges can be met with better use of technologies that already exist – particularly energy efficiency. Indeed, it is unlikely that technologies requiring radical jumps in technology and large infrastructures – from nuclear fusion to solar power beamed to earth from satellites – could be developed and deployed fast enough to make much of a difference in the dramatic shift away from fossil fuels that must be accomplished by 2050 if devastating climate disruption is to be avoided. Such radically new technologies may, however, be very important in the second half of the 21st century, when still-growing energy demand will have to be met with still-lower carbon emissions and may be encountering more problematic availability and security of adequate supplies of oil.

Nevertheless, to meet the challenges just described at reasonable cost, without incurring undue risks in other areas, will require major improvements in the cost and performance of energy technologies. In particular, the world urgently needs:

- New designs and materials that can drastically reduce energy use in buildings, industry, appliances, and vehicles
- Low-cost, non-polluting sources of light, heat, and electricity for the billions of people who still rely on wood, crop waste, and dung as their principal energy sources
- Fuel and vehicle systems that do not rely on oil, but can compete head-to-head with gasoline and diesel-powered vehicles
- Cost-effective energy storage for both vehicles and electric power systems
- Solar, wind, and other renewable energy sources at prices competitive with fossil fuels from a broad range of sites

- Competitive approaches to using fossil fuels while controlling emissions of carbon and other pollutants, such as carbon capture and sequestration (CCS)
- New nuclear designs that can provide cost-effective power and process heat with reduced safety, security, waste, and proliferation risks
- Smart grid technology that can manage electricity producers and users to efficiently meet demand, smoothly incorporating intermittent and distributed sources of energy and reducing unneeded uses when costs are high

The immense scale of the energy challenge and the heterogeneity in the energy sector make clear that no one technology will be enough: there is no silver bullet. Indeed, whether the technology in question is efficiency, solar, wind, nuclear, biofuels, or fossil fuel with carbon capture and sequestration, there are daunting obstacles to scaling up enough to be even 10% or 20% of the solution by 2050 (Pacala & Socolow, 2004). The world is likely to need many different approaches to energy efficiency and low-carbon energy generation, developed and deployed as fast and at as large a scale as it is practicable and cost-effective to do so. At the same time, neither the U.S. government nor anyone else knows today which technologies will prove to be most promising in 10 or 20 years. Hence, as described in Section 1.8 below, the U.S. government should invest in a broad portfolio of energy technologies, some of which will fail and some of which will succeed – just as an investor is wise to spread his or her resources across a broad portfolio of stocks.

1.5. Why Private Markets Are Essential But Insufficient

In the United States and most other countries, private firms are the main drivers of the economy, of innovation, and of energy production and use. The private sector will clearly play a central role in developing and deploying the new and improved energy technologies the world needs. There are several fundamental reasons, however, why private markets alone, without government support or regulation, cannot develop and deploy the technologies needed to meet the world's energy challenges. In the United States today, although there is general acceptance that the government has a positive role to play in supporting basic science and early-stage development, there is fierce debate over government's role at later stages of development and deployment of energy technologies. The available evidence makes a powerful case for government to work cooperatively with the private sector to overcome a series of market failures that would otherwise limit the pace of both invention and deployment of new and improved energy technologies.

First, the costs of many of the most critical of the world's energy challenges – emissions, the needs of the global poor, the national security impact of over-reliance on imported oil – are not appropriately priced in current markets. When oil, natural gas, and coal provide cheap energy and the negative environmental problems associated with their consumption are not reflected in prices, the private sector has only modest incentives to invest and innovate to develop or utilize alternatives.

Second, the benefits of energy research and development (R&D) to the entire private sector and to society at large are far greater than the benefits that would accrue to any particular company investing in such R&D, even if all the security, economic, and environmental issues were fully reflected in market prices. As research has clearly proven, companies (or countries) get only a part of the benefits of the R&D they invest in, and are able to free-ride on the R&D investments of others. The end result of this is that companies, overall, invest less in innovation than would be best for the interests of society as a whole.[3]

Third, a large part of the innovation potential lies with the end users of energy, who often have little incentive to pursue it. For most companies (and still more for most individuals) reducing their energy use is not a top priority; for many, energy costs are too small to warrant any significant efforts in energy innovation in the absence of government action. Moreover, many companies and individuals have limited knowledge about how much energy they use and how much could be saved – either with technologies already available or with new innovations.

Fourth, while the private sector is responsible for building and operating much of the energy system, this takes place in a context shaped by government support for and regulation of all of the associated infrastructure – from highways and gas stations to transmission lines and pipelines. These government policies must be shaped to maximize their overall benefit to society – but today, they are often shaped around the incumbent energy technologies on which society has been relying for decades.

Fifth, current market structures often slow energy innovation by creating a series of "valleys of death" for new technologies. For many technologies, there is a valley of death between the basic R&D for which government support is available and the full-scale development and deployment for which the risks are low enough for private financing at reasonable rates. In other cases, there is a valley of death between the small private investments that venture capital firms might be willing to make in the initial stages of development and the billion-dollar cost of building a large-scale demonstration plant that might convince larger investors such as private equity firms that the technology was ready for privately financed deployment. Government policies, including selective cost-sharing with industry, can help technologies get across these "valleys of death" – and by removing those obstacles to commercial success, leverage more private-sector investment (Narayanamurti et al., 2011).

Indeed, the economic literature emphasizes that it is necessary to use a combination of price and R&D policies to achieve the most cost-efficient transition to low-carbon energy systems with which to address the aforementioned market failures of the environmental externality and the spillovers from RD&D (Stavins, 2010). (For case studies on the role the U.S. government played in the long-term development of both shale gas and solar photovoltaics; see box on pp. 14–16.)

[3] Empirical evidence suggests that the social rates of return to R&D are substantially in excess of the private returns to R&D. While estimates of social rates of return to R&D in the United States range from 0 to 160%, those of private return range from 0 to 43% (Griliches, 1992; Margolis & Kammen, 1999).

Long Timescales and the Role of Government: The Stories of Shale Gas and Solar Photovoltaics

The history of global energy transitions going back to the Industrial Revolution in the mid-eighteenth century shows that they take place over multiple decades. It took more than a century for steam power from coal (which provided mechanical power and cheaper transportation) to go from 1% to 50% of the global energy system, and about 80 years for more modern energy sources – hydropower, gas, oil, nuclear, and modern renewables – to make up 50% of the global energy system, displacing coal-based steam (Gruebler, 2012).[4] Both of these transitions were driven by new applications (the many things made possible by steam engines, by electricity, and by the use of oil and gas), and the underlying technologies were initially relatively expensive. Replacing fossil fuels for power generation and transportation will require the development and deployment of cheaper alternative supply and end-use energy technologies. Many of these technologies do not offer performance improvements if externalities are not priced in, at least in their most important markets, which makes the role of governments, as we have already argued, even more important.

The history of two technologies largely developed in the United States that are having a more recent impact in the energy sector – shale gas and solar photovoltaics – is a testament to both the long timeframes involved in the development and deployment of new energy supply technologies and to the role the government played.

Shale gas has transformed the energy sector in the United States. It made up for less than 5% of all gas produced in the United States in 2005, rising rapidly to 11% in 2009, 23% in 2010, and 34% in 2012 (EIA, 2013). The existence of shale gas had been known for about 70 years, but no commercially viable technology was available to extract it from low-permeability shales.

The history of commercial shale gas extraction involved the development of hydrofracking, horizontal drilling, and microseismic monitoring technologies, and shows how the combination of private actors and the government over long periods of time has had a big impact on the U.S. energy system. The MIT *Future of Natural Gas* report (MIT, 2011) provides a detailed account of this history, starting with the authority that the Federal Energy Regulatory Commission (FERC) had to enable funding of R&D by industry, focused on the needs of the natural gas consumer. Using this authority, it funded the Gas Research Institute (GRI) in 1976 as a private nonprofit research management organization funded through a surcharge placed on interstate pipeline shipments of natural gas. Every 5 years, GRI submitted a comprehensive R&D plan to FERC for approval. The GRI was in essence the research arm of the gas industry, and, thanks to the surcharge in a regulated environment, which allowed pipeline companies to pass the cost to ratepayers, it received reasonably stable funding of over $200 million a year over extended periods.[5]

[4] As Gruebler (2012) notes, this second transition is far from complete, since approximately 2 billion people still have no access to modern energy carriers.

[5] The GRI reorganized and merged with the Institute for Gas Technology to become the Gas Technology Institute (GTI) in 2000, but its mission of serving natural gas consumers disappeared. The surcharge for the (then) GTI was ended in 2004. As the MIT (2011) report discusses, deregulation of the supply and interstate transportation of natural gas while keeping distribution and end-use markets regulated does not provide incentives for innovation.

Between the late 1970s, 1980s, and 1990s, DOE was also conducting gas research at a lower level of funding than the GRI. The two organizations engaged in joint portfolio planning to ensure that their programs were complementary and non-duplicative. The MIT report states (App. 8A, p. 5) that DOE programs "can be credited for assessing the resource base and the expansion of knowledge on natural fracture networks. Core and fractigraphic analysis carried out by DOE assisted in the deployment of massive hydraulic fracturing. The focus of the GRI program [funded by a FERC surcharge] was on the commercialization and deployment of technologies that were of interest to the industry, including new logging techniques, reservoir models and stimulation technologies."

Overall, an NRC report attributes the shale gas revolution to a large extent to DOE research, the long-term and stable, industry led model of research at the GRI, and an unconventional gas tax credit that was in place for almost two decades.[6] The drive, tenacity, and investments of industry pioneers like George Mitchell were also essential, since they led to the GRI's focus on shale gas and hydrofracking.

The history of solar photovoltaic (PV) power shows that a combination of public and private efforts over multiple decades has also had a dramatic impact on the improvement of solar PV.[7] Although the discovery of the photovoltaic effect dates back to the nineteenth century, the first solar cell using silicon was manufactured in 1954 by Daryl Chapin, Calvin Fuller, and Gerald Pearson at Bell Labs. The first important application of solar power came with the provision of power for space satellites in the late 1950s, an example of a niche market with tolerance for high prices driven by (in this case) government demand. In 1964, NASA launched the first Nimbus spacecraft, a satellite powered by a 470 W array (DOE, 2013c). In the 1970s, remote areas that were prohibitively expensive to connect to existing grids provided additional markets for what were very expensive solar panels – in 1975, solar PV panels still cost almost $100/W (Nemet, 2009). The private sector continued Bell Lab's push developing solar PV technologies with government support. For example, Spectrolab Inc. (a subsidiary of Boeing) came up with the use of screen printing during a contract with NASA's Lewis Research Center to set up a ground station in Western Africa. The oil crisis of the 1970s provided additional motivations to invest in solar PV RD&D. During this time, oil majors such as Exxon and Atlantic Richfield Oil Company (ARCO) were involved in solar PV research through SolarPower Corp. and ARCO Solar, respectively. ARCO Solar, for example, developed the use of aluminum-based pastes and was the first company to produce more than 1 MW of modules in one year (DOE, 2013c). President Carter's response to the oil crisis of the 1970s included the involvement of NASA's Jet Propulsion Laboratory in the development of terrestrial silicon PV modules by creating a procurement program that helped increase and establish their reliability and lifetime. The incentives and specifications provided by this procurement program played an

(*continued*)

[6] The focus of this piece is not about whether or not there are dangers involved with shale gas extraction, but rather on the role that the government and a close interaction of public and private entities had on the development of a technology having a large impact in the U.S. energy system.

[7] Some of the preliminary work on the history of solar PV innovation was developed by Laura Diaz Anadon working in collaboration with Christoph Wuestemeyer, master's student at MIT and research assistant at HKS in 2013.

important role in the solar PV cells we have today. Also during this time (in 1977) the U.S. government created the Solar Energy Research Institute, which then became National Renewable Energy Laboratory (NREL), to conduct R&D in solar energy. As described in Chapter 3, NREL had a major role in the development of First Solar, the largest thin film company in the world commercializing cadmium telluride (CdTe) cells. The costs of silicon-based PV cells have come down to $0.50/W in 2013 (which is closer to being competitive in large markets).[8] The research and incentives of the U.S. government and the work of the U.S. private sector over more than half a century have played an important role in this history of innovation.

1.6. Why Energy Innovation Is So Challenging

Unfortunately, low investment in innovation seems to be particularly severe in the energy sector. Those firms from the energy sector for which data are available spend a smaller portion of their revenues on innovation than do firms in a broad range of other industries (though, as the new data provided in this book make clear, past estimates have understated the full amount of private investment in energy innovation).[9] A look at the incentives for innovation in different industries makes clear why this is the case.

The primary role of private companies is to seek profits. They invest in R&D only when doing so appears to be the way to garner the best risk-adjusted rate of return for their shareholders. If a company is in the cell phone industry, for example, the pressure to innovate is extreme. Customers buy new phones frequently, choosing on the basis of features, style, and price. Competitors are developing and bringing to market new phones with new features so steadily that the entire life cycle of a particular phone model, from "cradle to grave," may be only a year or two. In the cell phone industry, a company that does not invest in innovation will soon be swept aside by its competitors. Innovation is the lifeblood of the firms in the industry.

Similarly, consider a large pharmaceutical firm. Such a firm would typically have many drugs to sell, but only a few that would bring in the majority of its profits. Those profits tend to decline greatly when the patent on a drug expires and other makers can compete with generic versions, driving prices down. Here, too, if the company is

[8] The 200-fold cost decrease between 1975 and 2013 is also due to significant government incentives for deployment in Japan, Germany, the United States, and Spain, among other countries. Some of these incentives have failed to deliver local manufacturing industries that their governments hoped for (Choi & Anadon, 2013) and many of these policies have had important unforeseen consequences, including expenditures that were far higher than expected (Hoppmann et al., 2013). We are not arguing that the economic returns to the countries supporting deployment policies are positive, but instead that the existence of these markets has contributed to the $50 billion market in solar PV which has enabled economies of scale.

[9] Utilities in particular, which include establishments engaged in the provision of electric power, natural gas, steam, water, and the removal of sewage, spend a very small fraction (0.1%) of net sales on R&D. Mining and extraction companies spend about 2% of sales on R&D. Manufacturers of motor vehicles, trailers, and parts invest 2.5% of sales in R&D. The pharmaceuticals, electronics, and computer equipment manufacturers all invest more than 10% of sales in R&D (NSF, 2008b). As discussed in Chapter 4, new data from our own survey and improved statistics from the National Science Foundation both suggest that private sector investment in energy innovation is larger than previously understood, but still substantially below the level that would likely be optimal for society as a whole.

not constantly investing in developing new drugs to provide a stream of profits in the future, it will soon be out-competed by companies that do make such investments.

By contrast, consider an electric utility that is in the business of selling electricity for slightly more than the cost of making it, primarily using large power plants that cost billions of dollars to build and that last for decades. Electricity is a commodity product – it does not come with special new features in the way a cell phone does. Any interruption of electricity supply could cost the utility millions of dollars a day. Its incentives are to focus on operating its existing facilities as cheaply and reliably as possible, and on buying new, reliable, and cheap facilities as needed. Neither the utility nor the companies that might hope to entice it to purchase a new power plant have much incentive to focus on innovation that goes beyond cost and reliability – except when government regulations and other government policies create such incentives. Moreover, the timescale of change is measured in decades rather than months. And there is little risk that other companies will take the utility's business if it fails to invest sufficiently in innovation.[10]

Much the same logic would apply to a large oil company. Such a company is in the business of producing oil and products made from oil at a somewhat lower cost than it can sell them. The business is based on oil and natural gas fields, exploration and extraction rigs, and refining plants whose development and construction cost billions of dollars, and which last for decades. As with electricity, the products are commodity products, much the same as what other companies are selling, not something like a cell phone or a cancer drug that can be sold for a higher price because it has features or capabilities that cannot be found elsewhere. An oil company has incentives to innovate in ways that allow it to find and produce more oil, or to produce oil and oil products more cheaply – but generally little incentive to invest in alternatives that today are more costly. (Because the large oil companies are wealthy firms, some of them have had the luxury of making some investments in new energy technologies such as biofuels.)

In addition, companies in the energy sector need to operate within an extensive enabling infrastructure. Pipelines, transmission lines, roads, and gas stations are essential components of today's energy system. In the future, other infrastructures will need to be developed to enable new technologies. For example, depending on what novel technologies end up contributing to reducing emissions and oil consumption in the transportation sector, different supporting infrastructures will need to be developed, such as power charging stations for electric vehicles, compressed natural gas (CNG)

[10] The telecommunications sector between the 1920s and the 1980s demonstrates that it is possible for a heavily regulated, integrated utility nonetheless to be a leader in innovation. In that case, however, the huge overall system was composed of many small parts – phones, switchboards, and the like – which could be switched out rapidly, and the company that had the monopoly was created with a strong culture of innovation. In addition, R&D and manufacturing were very closely linked, hinting at the importance for advancing innovation of such linkages between different stages of the value chain.

stations for CNG vehicles, or hydrogen pipelines for fuel cell vehicles. Because large-scale commercial success would require these infrastructures, which are not yet in place, the incentive for private sector innovation is less than it would otherwise be, though companies – from small start-up companies to electric cars from major automakers – are making investments to capture the niche markets and reputational benefits that are already available.

On the demand side there are also several barriers, albeit of a different scale and with different timelines. Taking residential buildings (which account for 21% of U.S. energy use) as an example, the reality is that most people simply do not spend the time and energy required to figure out what energy-efficiency measures would make sense for their home. Behavioral research has indicated that consumers do not know how much energy appliances or their homes use or where that energy comes from; assume new appliances are efficient; do not consider energy when they make purchasing decisions; focus on up-front costs rather than costs over time; and behave as though they had very high discount rates, making it more difficult for energy-saving investment to pay off (McKeown, 2007; Brown et al., 2009; Lutzenhiser, 2009). In addition, the building industry is highly fragmented, with a noticeable lack of energy efficiency knowledge in the workforce (Brown et al., 2008).

There are several fundamental reasons why innovation in energy technologies is particularly challenging:

- The large scale, high capital cost, and slow turnover time of many energy technologies, and the long timeframes over which their development takes place, slow innovation, create high barriers to entry, and limit private sector investment interest in energy innovation. In addition, the strong economies of scale in many energy supply technologies mean that commercially viable facilities are large and expensive – and in many cases, there is substantial learning-by-doing, meaning that the first few plants of a new type tend to be more expensive than that technology will be once the kinks are worked out. Financing these more expensive first plants poses a major barrier to development and deployment of some energy technologies.
- There are financial pressures on corporations to allocate resources to investments with more predictable benefits for the short-term bottom line than RD&D. Established firms tend to focus resources preferentially on existing competencies and incremental improvement, and away from innovative alternatives that could make their present products obsolete.
- Positive feedback loops within a system, network, or infrastructure can "lock-in" existing technological solutions, making incumbent technologies difficult to displace and locking-out the integration of alternative technologies. There is a "chicken and egg" problem associated with the large investments in new infrastructure needed to make some of the new technologies effective at scale.
- Energy, by nature, is a commodity subject to price fluctuations that impose large uncertainties on the returns to be expected from investment in innovation (Anadon & Holdren, 2009).

The energy industry, with its focus on commodities produced using expensive facilities designed to last for decades, is inevitably an industry inclined toward a slow pace of technological change. Although venture capital (VC) and private equity investors have increased their support for companies in the "clean tech" sector from $600 million in 2004 to $5.9 billion in 2011 and $4.3 billion in 2012 in the United States (BNEF, 2013) – down from a peak of $7 billion in 2008 – it is not clear that the VC model will be as successful commercializing supply-side energy technologies as it was with information technology (IT) companies given the large capital requirements of large energy supply projects (Ghosh & Nanda, 2010). Even in that sector a large part of the innovations in the early decades came from near monopolies like AT&T that had a vertically integrated structure from suppliers to manufacturing. However, the VC model in the energy sector may work in customer side areas such as solid state lighting and appropriate transportation technologies.

Private markets will be essential drivers of the needed energy innovations, but they will have to be private markets with new funds, regulations, and other incentives established by government policies.

1.7. The Energy Innovation System

How does energy innovation work? How can government policy best accelerate the invention and commercialization of new and improved energy technologies? These are hard questions. Many uncertainties remain about the answers. But a good deal is known about how the energy innovation system behaves.

Energy innovation can best be thought of as an ecosystem (Gruebler et al., 2012). The ecosystem will be productive only if it has enough nutrients (funds, ideas, and appropriately trained and motivated people), and if all the various elements of the ecosystem, namely the actors, institutions, and linkages between them, are in balance.[11] For the ecosystem to be productive, innovation institutions (which include institutions responsible for crafting policies) must strengthen or create all of the elements of this ecosystem in a balanced way. This book attempts to provide recommendations about how to stimulate innovation by different actors, and how to strengthen the linkages between major innovation, the actors, and institutions.

Innovation has often been described as a linear process in which technologies move from basic research to applied research, applied research to development, development to large-scale demonstration, and demonstration to commercial deployment and diffusion. In reality, it is much more like a food web in an ecosystem, with technologies and resources flowing both forward and backward, as shown in Figure 1.2. The users of a new technology may come up with ideas that lead to modifications

[11] Key institutions in the ecosystem include businesses, industry associations, institutions facilitating technology transfer, organizations performing R&D (e.g., universities, national laboratories, etc.), and entities "guiding" innovation (institutions creating policies and regulation, such as the Department of Energy, Congress, and the Environmental Protection Agency). Key actors include entrepreneurs, scientists, consumers, and activists, among others. The ecosystem is also shaped by established habits, practices, and norms (Gruebler et al., 2012).

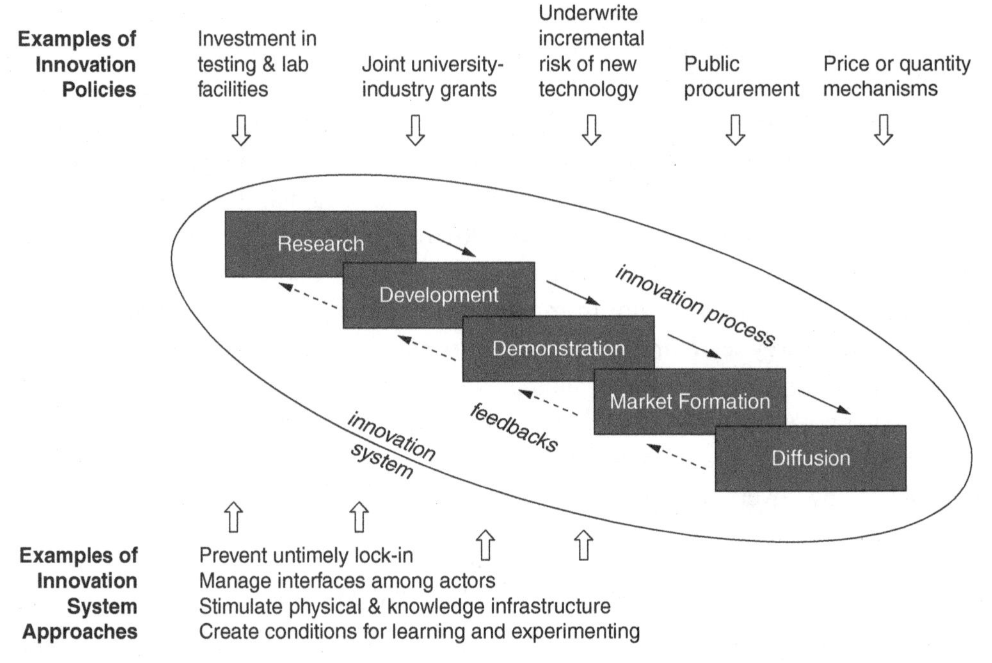

Figure 1.2. Schematic of stages and linkages of the energy innovation system (GEA, 2012).

in the next version, or even to new basic or applied research; researchers may come up with ideas that will be directly incorporated into the next generation of deployed technology (Gruebler et al., 2012). Indeed, even the oft-repeated distinction between basic and applied research breaks down on closer examination of this ecosystem (Narayanamurti, Odumosu, & Vinsel, 2013).

Major gaps at any point can disrupt the entire system, just as removing a top predator or a crucial food species can have cascading effects throughout an ecosystem. For example, if private firms see no way to finance the risk of a commercial-scale demonstration of a new technology, and there is little prospect of government support for such a demonstration, they will have little incentive to invest in the early-stage research and development that might lead to such a demonstration – even if there are reasonable prospects that the technology could someday be commercially successful after the demonstration hurdle had been overcome.

Government policies have an important role in strengthening different parts of the innovation ecosystem. First, as noted earlier, they should strive to overcome barriers that prevent private markets from reflecting the full value of energy innovation to society – from "valleys of death" to misplaced incentives. Second, they must ensure that there is a sufficient supply of trained scientists and engineers. Third, strong cultures of innovation that foster a focus on new ideas and a willingness to take risks are crucial, particularly in the innovation institutions supported by U.S. government funds (Narayanamurti, Anadon, & Sagar, 2009). Fourth, policies can contribute to the creation of environments and mechanisms to exchange ideas and forge agreements among different actors in the system opportunities for strengthening the linkages. Start-up firms, large energy companies, research universities, venture capital, private equity, investment banks, government agencies, non-government organizations, legislatures, the media, and the public all have important roles to play.

Policymakers must also recognize that there are important feedbacks among all the elements of the energy innovation ecosystem. For example, policies that create market demand for the deployment of clean energy systems (such as a substantial carbon price or renewable portfolio standards) will lead to increased private R&D and the formation of new start-ups seeking to profit from meeting the new demands – although quantifying this effect and determining what policies are most efficient at stimulating it are far from straightforward. In Chapter 4 we make a first attempt at studying how important different government activities are in stimulating innovation, by asking business establishments participating in our energy innovation survey what government policies affect their decisions about energy RD&D.

1.8. CASCADES: Criteria for an Effective Energy Innovation Policy

It is not easy to develop policies that will succeed in accelerating the pace of energy technology innovation. Our research has convinced us that an effective energy innovation approach must meet the following criteria, which collectively we call the CASCADES approach. An effective approach must have all of the attributes described in the following subsections.

1.8.1. Comprehensive

Given the complex energy innovation ecosystem just described, it is not enough just to provide some research support and hope the result will be commercialized. As discussed earlier, even large investments in R&D without substantial policies to support demonstration and broad-scale deployment are not likely to be enough. Nor is it enough to put in place a requirement to deploy a certain amount of renewable energy, or financial incentives for deployment, and hope cost-effective technologies will be developed as a result. Investing to develop cost-effective approaches to capturing and storing carbon from burning fossil fuels, for example, will have little benefit if there is no limit or price on emitting carbon – there will be no financial incentive to deploy the carbon-capture technologies at a scale large enough to make a difference. Effective energy innovation policies must be *comprehensive* enough to overcome all the market failures and barriers to deployment that exist throughout the innovation life cycle, integrating R&D support with appropriate policies related to technology demonstrations, niche deployments, and broad-scale market adoption. This means investing in a broad portfolio of technological stages of energy technologies, from basic science that could lead to long-term future breakthroughs to commercial-scale demonstration and initial deployment of technologies that are almost ready for commercial deployment.

1.8.2. Adaptive

Energy innovation policies must also be *adaptive*, learning from experience and adjusting as they go. Some technological pathways or approaches to innovation will turn out to work well; others will not. Some will provide solutions the marketplace will embrace, while others will be left behind by changing market conditions and more attractive alternatives. Energy innovation strategies should be designed explicitly to incorporate learning from what works and what does not and to adapt to a changing economic and technological landscape over time. Species that fail to adapt to a changing environment will become extinct; energy innovation policies that fail to incorporate ongoing learning and adaptation will waste the taxpayers' money.

A willingness to take risks – to accept that some projects will fail – is a crucial part of such an adaptive approach. Venture capital firms often expect that nine out of every ten projects they fund will fail – while the tenth will pay off many times over, providing the returns that keep the firm running. Indeed, learning from a "failed" project may provide new insights about what works and what does not that further the success of the overall adaptive effort. A central theme of this book is that rapid innovation requires an acceptance of risk. Policymakers in Washington must be willing to accept that a significant portion of projects will fail, avoid making a political *cause célèbre* out of each perceived failure, and be willing to continue support for innovation efforts even in the face of such failures. Knowledge gained through this experience provides the ability to terminate projects gracefully and helps to identify new ventures that might be of interest for the United States' energy future.

1.8.3. Sustainable

Any innovation strategy – but particularly one focused on energy technologies, where the underlying capital stock turns over on a timescale of decades – has to be *sustainable*, built to last over time. Investing in one technology today and dropping it for another a year from now will bear little fruit. As we describe in Chapter 2, U.S. energy R&D investments have suffered from wild oscillations in focus and approach, with the amount of funds available for particular technologies way up one year and way down the next. The fate of hydrogen-powered vehicles – heavily promoted in the George W. Bush administration, rejected in the Obama administration – is just one example of this on-again off-again approach (regardless of what one thinks of the merits of the two administrations' approaches to that particular technological choice).

A focus on long-term sustainability – including a search for broad bipartisan support for major initiatives – has to be built into U.S. energy innovation strategy. A technology demonstration approach that required Congress to decide to appropriate another billion dollars for a particular technology every year for 15 years, for example, would be unlikely to succeed, while options that are self-financing from revenue streams outside the typical appropriations process might well be able to sustain themselves. In the current polarized political environment in the United States, building sustainable bipartisan support for an approach may be the biggest challenge that energy innovation strategies face.

1.8.4. Cost-effective

Energy technology innovation policies should seek to achieve energy technology innovation goals at minimum cost – not just to the government, but also to the society as a whole. In some cases, it may be most cost-effective for government policies to structure market incentives and let the private sector respond to them; in other cases, it may be that government investments in RD&D designed to drive the cost of a particular technology down or its performance up may enable widespread deployment at lower cost than would be the case if the government invested primarily in subsidizing deployment of the technologies already available. Such choices require expert judgment, and there will always be a need for balance – but the goal of that balancing should be to be *cost-effective*, to provide the most progress toward established energy technology innovation goals per dollar spent.

1.8.5. Agile

Although energy technology innovation strategies must be sustainable and focused on the long term, they must also be *agile* enough to respond to developments as they arise. If a new technology, an economic shift, or a newly discovered resource creates a new opportunity – or closes off an old one – energy technology strategies must be able to move quickly to seize the moment. This need for agility must, of course, be balanced against the need for sustained investment to make progress in technologies. But a lack of agility – for example, continuing to pour money into a technology long after it is clear there is little prospect for economic viability – can be just as damaging as an investment focus that flits from one technology to another too quickly.

1.8.6. Diversified

Effective energy innovation policies must not only cover a broad array of stages of technological development but also a *diversified* portfolio of technologies. As noted earlier, no one technology is going to be enough to meet the complex energy challenges the United States faces – and which technologies will prove to be most promising remains highly uncertain. Hence, the U.S. government must invest in a broad portfolio of energy technologies – just as a stock investor is wise to put his or her money in a broad portfolio of stocks. As with a stock investor, however, this does not mean investing in promising and unpromising options alike – technical judgments will have to be made, but they should keep open a variety of options, and avoid excluding options before their potential is fully explored. This emphasis on a portfolio approach for technologies, stages of research, and partnership mechanisms (among others) is a central theme throughout this book.

1.8.7. Equitable

A strong energy innovation policy should also be *equitable* – treating technologies, companies, and regions of the country without arbitrary favoritism. In the United States and in many other countries, there has been too much of lobbyists and lawmakers carving out specific subsidies for specific industries and companies. Sprawling energy legislation sometimes becomes festooned with provisions intended to benefit particular special interests; one such bill was derided as the "no lobbyist left behind act."

Instead, to the extent practicable, the U.S. government should adopt a technology-neutral set of policies designed to correct market failures by providing different types of support at different stages of the technology life cycle. Rather than government legislating how much of which technologies should be deployed, government should establish goals – for example, for certain levels of emission of carbon or of pollutants such as fine particulates – structure markets so that firms have an incentive to meet those goals, and let markets decide which mix of technologies is best for the job.

Such fairness is difficult to achieve. Each technology has particular needs that are less relevant for the others. Policy on pipelines is a key policy issue for the oil and gas industries, but irrelevant to most other energy industries. For nuclear energy, measures to reduce the risk to investors (e.g., through long-term power purchase agreements, loan guarantees, or guaranteed rates of return in regulated markets) are among the most crucial forms of support, because the substantial risks, delays, and cost over-runs in nuclear projects drive up financing costs. Solar facilities generally pose much lower risks, but nevertheless have high costs – so production tax credits are a particularly important type of support for the solar industry. And each industry represents an entrenched set of interests with its legislative backers, making large-scale change in past approaches difficult. Nevertheless, it is crucial to try, as much as possible, to design technology-neutral approaches; an approach to driving down risks for nuclear investors, for example, should also be available for other technologies that offer similar societal benefits and face similar problems with risk driving financing costs.

1.8.8. Strategic

Finally, an effective energy innovation policy must be *strategic* – it must be focused on clearly identified goals, and on technologies and approaches that offer significant promise of contributing to those goals. If, for example, a particular technology "works" but has little prospect of reaching the scale of deployment necessary to have a substantial impact on the major U.S. energy innovation goals, it may be time to consider phasing out support for that technology – or concentrating support only on exploring whether the roadblocks to large-scale deployment can be overcome. Certainly technologies should not be subsidized if they actively undermine the strategic goals of the energy innovation strategy – yet around the world today, almost twice as much of the world's carbon emissions receive government subsidies (15%) as are subject to any form of carbon price (8%), and the subsidies are higher than the prices, amounting to the equivalent of more than $100 per ton of carbon emitted; in total, fossil fuels received over $500 billion a year in subsidies worldwide, some six times the government support provided for renewable energy in that year (IEA, 2013, p. 11).[12]

What should the strategic goals of the energy innovation strategy be? We propose that they focus on the key items in the opening question in this book: providing adequate supplies of reliable, affordable energy to all, without causing catastrophic climate change or other environmental disasters. In the context of deployment of energy technologies within the United States, discussed in Chapter 2, the key criteria included in our economic modeling included overall carbon emissions (or the cost of reducing carbon emissions in scenarios where emissions were capped); oil dependence; and the cost of meeting energy demand.

1.8.9. Meeting the Criteria: Synergies and Tradeoffs

Of course, these criteria interact with each other in complex ways. In many areas there are strong synergies: an approach that meets the equitability criterion is more likely to meet the cost-effectiveness criterion, for example. In other areas there are tradeoffs – between the sustained investments needed to make progress in energy innovation, for example, and the agility required to seize new opportunities. Overall, however, we believe that only energy innovation strategies that offer a balanced approach to fulfilling the CASCADES criteria will maximize the chance that U.S. government policy can create cascades of beneficial energy innovations stretching into the future.

[12] Of course, these fossil fuel subsidies are themselves intended to achieve other government goals, primarily social goals such as ensuring that the poor have access to energy. But studies have shown that they are extremely inefficient means of reaching these goals; by one estimate, in cases where the subsidy is intended to help the poor, "only 8% of the subsidy granted typically reaches the poorest income group" (IEA, 2013, p. 49). Nevertheless, once such subsidies are in place, reducing the prices for commodities such as heating oil, gasoline, or electricity, people resist giving them up, making them politically extremely difficult to change. They thus fail the strategic focus criterion, the cost-effectiveness criterion, the agility criterion, and the equitability criterion described here at the same time.

1.9. Plan of the Book

This book is structured to offer insights into how the U.S. government could best accelerate the entire energy innovation ecosystem, leveraging the many different innovators in both the public and private sectors. One chapter is devoted to each of the four large questions posed at the outset of this chapter.

First, Chapter 2 discusses how much the U.S. government should spend on energy RD&D, and on what. To make recommendations for an expanded and retargeted U.S. federal investment in energy RD&D, we sought to understand what improvements in cost and performance might result from such investments, and what real benefits those improvements would have for addressing climate change or improving U.S. energy security, under different deployment policy scenarios.

To do this, we began with expert elicitations in a range of key energy technologies, which assessed how much RD&D funding the experts believed each technology required, how those funds should be allocated, and how much progress they believed would result from their recommended RD&D budgets – that is, how their recommended budgets would shift the spectrum of possible cost and performance that technology might achieve by 2030. We also asked the experts to estimate the uncertainty in these projected outcomes, priming them with information designed to reduce the overconfidence bias that many experts have (Tversky & Kahneman, 1974; Brenner, Koehler, Liberman, & Tversky, 1996). We supplemented these expert elicitations with in-depth interviews with individuals who are actively involved in making funding decisions related to that technology.

We then used the MARKAL (MARKet ALlocation) economic model to explore how the improved cost and performance projected by the experts – with their associated uncertainties – might affect commercial deployment of these technologies in a range of different policy scenarios, and how those deployments might affect goals such as reducing greenhouse gas emissions or dependence on imported oil. Finally, based on the expert surveys and modeling, Chapter 2 offers recommendations for U.S. energy RD&D investments in a range of energy technologies.

Chapter 3 then discusses our second question: How can the U.S. government manage its energy RD&D investments and the institutions through which they flow to get the most beneficial energy innovation per dollar invested? In Chapter 3, we explore this question through case studies of both public and private innovation institutions, including the National Renewable Energy Laboratory (NREL), funded and directed by DOE, the Semiconductor Research Corporation, a private consortium established to accelerate innovation in semiconductor manufacturing, and the Energy Biosciences Institute (EBI), which is an example of university–industry partnership focused on biofuels. Based on these case studies, on a broad literature on innovation, and on the experience of one of the authors of this chapter (Narayanamurti, who has led

R&D initiatives at Bell Labs, at Sandia National Laboratory, and at major research universities), we offer a range of recommendations for strengthening the energy innovation efforts of the national laboratories and other DOE-sponsored innovation institutions.

Chapter 4 then addresses our next large question: How can the U.S. government best cooperate with the private sector and structure incentives for private sector energy innovation? Answering this question requires an understanding of the current state of affairs in private sector energy innovation, and the current approaches the U.S. government takes to funding RD&D through private firms rather than national laboratories, research universities, or other nonprofit institutions. Hence, the chapter begins with data from a pilot-scale survey of private sector energy innovation in the United States, which significantly expands the data available on the scope of private sector energy innovation and the incentives that are most effective in driving it. The chapter then describes the government's approaches to channeling government energy RD&D investments through private firms, which largely focus on types of agreements that DOE characterizes as "grants" and "cooperative agreements," and how these partnerships could be managed more effectively. In addition, the chapter outlines our recommendations for a strengthened approach to managing large-scale commercial demonstrations of new technologies, pulling together insights from both government and private sector perspectives. These recommendations are based in part on the results of a high-level executive session held in December 2010, and include a specific discussion of institutional approaches.

Chapter 5 takes on our fourth large question: How the U.S. government can best manage both competition and cooperation with foreign countries in energy technology innovation. Here, too, an understanding of the problem and potential solutions requires an understanding of the current state of affairs – which is difficult, because international cooperation and competition on energy technology innovation is vast, diverse, often commercially proprietary, and rapidly changing. We offer new data and analysis on the investments emerging economies are making in energy technology innovation; on international energy technology cooperation between governments, corporations, and individual scientists; and on the diverse set of approaches the U.S. government has historically taken to international energy RD&D cooperation. Building on those analyses, Chapter 5 offers a new approach to identifying and promoting strategic opportunities for international cooperation, while allowing the origination of projects also at the agency and national laboratory level.

Finally, in Chapter 6, we summarize our arguments and recommendations and assess issues that cut across all of these categories. We believe that in each of these areas, this book makes a substantial contribution to the literature on energy technology innovation, broad and diverse though that literature already is.

1.9.1. Limitations

Of course, there are many questions this book leaves unanswered. The scope of the energy challenges is huge, and our research team was small. Hence, we were not able to include several important energy technologies in our expert surveys, or to do more than an initial pilot-scale survey of the private sector energy innovation, because of resource limitations. Our economic modeling and resulting energy RD&D funding recommendations are based only on the benefits of technologies developed in the United States for deployment within the United States. In reality, of course, there will be technologies developed in other countries which will be deployed in the United States (which might then reduce the need for the United States to invest in them, except from a competitive perspective), and technologies developed in the United States that will be deployed globally (which mean our modeling, reflecting only one small part of the market, would greatly understate the benefit of investing in them, either in economic terms or in terms of climate objectives). Our analysis does not explore the role that local and state governments play in promoting energy innovation – which can in some cases be substantial (for example in incentivizing deployment through policies such as tax credits and mandates). Even within the U.S. federal government, our analysis does not cover all of the agencies involved in energy innovation. For example, we did not study in depth how the Department of Defense, the nation's largest energy buyer, may be able to create niche markets for energy technologies through procurement – a topic of recent interest (see, e.g., John et al., 2010). Nor does this book explore the role of behavioral change in addressing our energy problem. Behavioral change is of direct relevance to this book, because changing attitudes impact the acceleration or limitations of deploying new energy technologies, and may impact the kinds of technologies that can be developed. Furthermore, behavioral change (in particular, changes in consumption patterns) has important implications for future energy needs, and therefore energy technology innovation.

At the same time, the methodologies used in this book have their own limitations. First and perhaps most important, predicting the outcome of RD&D is very difficult, and past experience suggests a high probability that the consensus of experts on what is promising will be wrong in some areas. Even though our expert consultation also asked experts to assess the uncertainty in their predictions and think about technology improvements that have low probabilities of occurring, the uncertainties are surely larger in some cases than the experts estimated, and some things will happen that the experts could not or did not predict.

Second, even though we considered the role of the private sector as a source of innovation, more research is needed to understand how government policies affect private innovation. How do policies such as carbon prices, efficiency standards, or requirements to deploy certain quantities of renewable energy capacity motivate or discourage private sector investment in improving these technologies – and how will those private sector investments affect the pace of future technological change? How do government energy RD&D investments affect private entities' RD&D investments? Do government investments encourage additional private investments in those areas ("crowding in") or lead to lower private investments ("crowding out"), or do both of these effects occur to differing degrees in differing circumstances? What are the opportunity costs from the government investing in energy RD&D and that money not being spent elsewhere in the economy?

Like all modeling of phenomena as complex as future economic, energy, and technology trajectories, our modeling of the economic and environmental benefits of federal energy RD&D investments has its limitations. Not only was our modeling limited to the United States, but, like other modeling in this area, it did not include a fundamental but difficult-to-model effect: if it becomes expensive to dump carbon into the atmosphere, companies will presumably invest more in finding ways to avoid doing so, and those investments will accelerate the pace of technological change. Moreover, the model we used incorporated a variety of constraints on the pace at which deployment of various technologies could expand, even once they became cost-competitive; these model constraints may turn out to be either tighter than is realistic (meaning that technologies are able to expand more rapidly than the model projects) or looser than is realistic (meaning that the non-cost difficulties in achieving widespread deployment are even greater than those incorporated in the model).

At the same time, our analysis of international cooperation activities is based on aggregated and often incomplete datasets with limited information about the objectives, reasons, or detailed descriptions of the actual activities within each broad category. Furthermore, we did not interview all of the agencies involved in international cooperation on energy technology. Finally, more work is required to understand the motives, needs, and mechanisms for international cooperation on energy technology innovation in other countries, and our current analysis of international cooperation activities without U.S. involvement is limited. Detailed case studies would be necessary to understand which international cooperation activities provide the most effective support for U.S. companies and the most effective long-term job creation within the United States.

Despite these limitations, we believe the overall picture we paint in this book is sound, and firmly grounded in the data and analysis we present. Fundamentally, to meet the vast energy challenges the United States and the world face will require a dramatic acceleration in the pace at which new and improved energy technologies are researched, developed, demonstrated, and deployed. The U.S. government needs an energy innovation strategy matched to the scope and urgency of the energy challenges and energy opportunities this nation faces. Such a strategy would include a substantially increased U.S. government investment in energy innovation, structured in a way that meets the CASCADES criteria, along with a broad range of policies to overcome market barriers to both innovation and deployment. By following the approaches outlined in this book, the United States can take a major step toward meeting the energy technology challenges of the coming decades and gaining an appropriate share of the huge energy technology markets of the future.

References

Acemoglu, D., & Robinson, J. (2012). *Why Nations Fail: The Origins of Power, Prosperity, and Poverty*. New York: Crown.

AEP. (2011). "AEP places carbon capture commercialization on hold, citing uncertain status of climate policy, weak economy." American Electric Power. 14 July. Columbus, Ohio, United States. Retrieved from http://www.aep.com/newsroom/newsreleases/?id=1704 (Accessed March 3, 2014).

American Energy InnovationCouncil. (2010). *A Business Plan for America's Energy Future*. Full Report. June. Washington, D.C.: Energy Innovation Council.

American Enterprise Institute, Brookings Institution, & Breakthrough Institute. (2010). *Post Partisan Power: How a Limited Approach to Energy Innovation Can Deliver Cheap Energy, Economic Productivity, and National Prosperity*. October. Washington, D.C.: American Enterprise Institute, Brookings Institution, & Breakthrough Institute.

Anadon, L.D., Bunn, M., Chan, G., Chan, M., Jones, C., Kempener, R., Lee, A., Logar, N., & Narayanamurti, V. (2011). *Transforming U.S. Energy Innovation*. November. Energy Technology Innovation Policy. Cambridge, Mass.: Belfer Center for Science and International Affairs, John F. Kennedy School of Government, Harvard University.

Anadon, L.D., & Holdren, J.P. (2009). "Policy for energy-technology innovation." In *Acting in Time on Energy Policy*, edited by Kelly Sims Gallagher, pp. 89–127. Washington, D.C,: Brookings Institution Press.

ARPA-E. (2010). "Advanced Research Projects Agency – Energy." Retrieved from http://energy.gov/articles/arpa-e-fy2010-annual-report-highlights-transformational-projects-agency-s-establishment/ (Accessed November 1, 2013).

BBC. (2010). "NATO tankers torched in Pakistan." October 1, 2010. Retrieved from http://www.bbc.co.uk/news/world-south-asia-11450011 (Accessed November 1, 2013).

BNEF. (2013). "New Energy Finance Desktop." *Bloomberg New Energy Finance*. Retrieved from http://www.newenergyfinance.com (Accessed November 1, 2013).

Brenner, L.A., Koehler, D.J., Liberman, V., & Tversky, A. (1996, March). "Overconfidence in probability and frequency judgments: A critical examination." *Organizational Behavior and Human Decision-Making Processes*, 65(3):212–19.

Brown, M.A., Chandler, J., Lapsa, M.V., & Moonis, A. (2009). *Making Homes Part of the Climate Solution: Policy Options to Promote Energy Efficiency*. U.S. Climate Change Technology Program. Oak Ridge, Tenn.: Oak Ridge National Laboratory.

Brown, M.A., Chandler, S., Lapsa, M.V., & Sovacool, B.K. (2008). *Carbon Lock-In: Barriers to the Deployment of Climate Change Mitigation Technologies*. Oak Ridge, Tenn.: Oak Ridge National Laboratory.

Choi, H., & Anadon, L.D. (2014). "The role of the complementary sector and its relationship with network formation and government policies in emerging sectors: The case of solar photovoltaics between 2001 and 2009." *Technology Forecasting and Social Change.* 82:80–94.

CNA. (2007). *National Security and the Threat of Climate Change.* Alexandria, Va.: Center for Naval Analyses.

CNA. (2009). *Powering America's Defense: Energy and the Risks to National Security. Report from the CNA's Military Advisory Board. May.* Alexandria, Va.: Center for Naval Analyses.

Daniel, L. (2010). "New office aims to reduce military's fuel usage." American Forces Press Service. Retrieved from www.defense.gov/newsarticle.aspx?id=60131 (Accessed July 22, 2010).

Defense Science Board. (2008). *More Fight – Less Fuel.* Office of the Under Secretary for Acquisition Technology and Logistics. February. Washington, D.C.: U.S. Department of Defense.

Deutch, J.M. (2011). *The Crisis in Energy Policy*, Cambridge, Mass.: Harvard University Press.

DOE. (2010a). Energy Frontier Research Centers. Main Website. Office of Science. U.S. Department of Energy. Retrieved from http://science.energy.gov/bes/efrc/ (Accessed March 3, 2014).

DOE. (2010b). Energy Innovation Hubs. Main Website. U.S. Department of Energy. Retrieved from http://www.energy.gov/science-innovation/innovation/hubs (Accessed March 3, 2014).

DOE. (2011a). *Department of Energy Quadrennial Technology Review Framing Document.* Washington, D.C.: U.S. Department of Energy. Retrieved from http://energy.gov/sites/prod/files/edg/qtr/documents/DOE-QTR_Framing.pdf (Accessed November 1, 2013).

DOE. (2011b). *Report of the Quadrennial Technology Review.* Washington, D.C.: U.S. Department of Energy September. Retrieved from http://energy.gov/sites/prod/files/QTR_report.pdf (Accessed November 1, 2013).

DOE. (2012). *Report on the First Quadrennial Technological Review: Technology Assessments.* DOE/S-0002. Washington, D.C. Retrieved from http://energy.gov/downloads/quadrennial-technology-review-august-2012 (Accessed November 1, 2013).

DOE. (2013a). Energy Frontier Research Centers. Main Website. Office of Science. U.S. Department of Energy. Retrieved from http://science.energy.gov/bes/efrc/ (Accessed November 1, 2013).

DOE. (2013b). Energy Innovation Hubs. Main Website. U.S. Department of Energy. Retrieved from http://www.energy.gov/hubs/ (Accessed November 1, 2013).

DOE. (2013c). "The history of solar PV." U.S. Department of Energy, Office of Energy Efficiency and Renewable Energy. Washington, D.C. Retrieved from http://www1.eere.energy.gov/solar/pdfs/solar_timeline.pdf (Accessed November 1, 2013).

Easterling, W.E., Aggarwal, P.K., Batima, P., Brander, K.M., Erda, L., Howden, S.M., Kirilenko, A., Morton, J., Soussana, J.-F., Schmidhuber, J., & Tubiello, F.N. (2007). "Food, fibre and forest products." In *Climate Change 2007: Impacts, Adaptation and Vulnerability.* Contribution of Working Group II to the Fourth Assessment Report of the Intergovernmental Panel on Climate Change, edited by M.L. Parry, O.F. Canziani, J.P. Palutikof, P.J. van der Linden, and C.E. Hanson, pp. 273–313. Cambridge: Cambridge University Press.

EIA. (2009). *International Energy Outlook 2009* [Homepage of U.S. Energy Information Administration. Department of Energy]. Retrieved from http://arsiv.setav.org/ups/dosya/25025.pdf (Accessed March 3, 2014).

EIA. (2013). *Annual Energy Outlook*, AEO2013 Early Release Overview. Energy Information Administration, Washington, D.C.: U.S. Department of Energy.

European Commission. (2006). *World Energy Technology Outlook – 2050 (WETO – H2)*, Brussels, Belgium: European Commission. Directorate General for Research.

G8+5 Academies. (2009, May). *G8+5 Academies' Joint Statement: Climate Change and the Transformation of Energy Technologies for a Low-Carbon Future*. Washington, D.C.: National Academies Press.

Gallagher, K.S. (2009). "Acting in Time on Climate Policy." In *Acting in Time on Energy Policy*. Washington, D.C.: Brookings Institution Press.

Gallagher, K.S., & Anadon, L.D. (2010). DOE Budget Authority for Energy Research, Development, & Demonstration Database. Retrieved from http://belfercenter.ksg.harvard.edu/publication/20013/ (March 3, 2014).

Ghosh, S., & Nanda, R. (2010). "Venture capital investment in the clean energy sector." Harvard Business School Entrepreneurial Management Working Paper No. 11–020. Retrieved from http://www.hbs.edu/faculty/Publication20Files/11-020.pdf (Accessed November 1, 2013).

Griliches, Z. (1992). "The search for R&D spillovers." *Scandinavian Journal of Economics*, 94:S29–S47.

Gruebler, A. (2012). "Grand designs: Historical patterns and future scenarios of energy technological change. Historical case studies of energy technology innovation." Supplement to Johansson, T.B., Patwardhan A., Nakicenovic N., & Gomez-Cheverri L., eds., *Global Energy Assessment – Toward a Sustainable Future*. Cambridge: Cambridge University Press, and Laxenburg, Austria: The International Institute for Applied Systems Analysis. Retrieved from http://www.iiasa.ac.at/web/home/research/researchPrograms/TransitionstoNewTechnologies/01_Grubler_Transitions__WEB.pdf (Accessed November 1, 2013).

Gruebler, A., Nakicenovic, N., & Victor, D.G. (1999). "Dynamics of energy technologies and global change." *Energy Policy*, 27:247–80.

Gruebler, A., Aguayo, F., Gallagher, K., Hekkert, M., Jiang, K, Mytelka, L., Neij, L., Nemet, G., Wilson, C., et al. (2012). "Policies for the energy technology innovation system (ETIS)." In Johansson, T.B., Patwardhan A., Nakicenovic N., & Gomez-Cheverri L., eds., *Global Energy Assessment – Toward a Sustainable Future*, pp. 1665–1744. Cambridge: Cambridge University Press, and Laxenburg, Austria: The International Institute for Applied Systems Analysis.

Henderson, R.M., & Newell, R.G., eds. (2011). *Accelerating Energy Innovation: Insights from Multiple Sectors*. Chicago: University of Chicago Press.

Hoppmann, J., Peters, M., Schneider, M., & Hoffmann, V.H. (2013). "The two faces of market support – How deployment policies affect technological exploration and exploitation in the solar photovoltaic industry." *Research Policy*, 42(4):989–1003.

IEA. (2009). *World Energy Outlook 2009*. Paris: International Energy Agency, Organization for Economic Cooperation and Development, OECD.

IEA. (2010a). *Energy Technology Perspectives 2010: Scenarios & Strategies to 2050*. Paris: International Energy Agency, Organization for Economic Cooperation and Development, OECD.

IEA. (2010c). *World Energy Outlook 2010*. Paris: International Energy Agency, Organization for Economic Cooperation and Development, OECD.

IEA. (2012). *World Energy Outlook 2012*. Paris: International Energy Agency, Organization for Economic Cooperation and Development, OECD.

IEA. (2013). *Redrawing the Energy-Climate Map*. Paris: International Energy Agency, Organization for Economic Cooperation and Development, OECD.

Inhofe, J. (2012). *The Greatest Hoax: How the Global Warming Conspiracy Threatens Our Future*. Washington, D.C.: WND Books.

IPCC. (2012). Managing the Risks of Extreme Events and Disasters to Advance Climate Change Adaptation. Cambridge: Cambridge University Press.

John, A., Sarewitz, D., Weiss, C., & Bonvillian, W. (2010). "A new strategy for energy innovation." *Nature*, 466:316–317.

Lester, R.K. (2009). "America's energy innovation problem (and how to fix it)." MIT-IPC Energy Innovation Working Paper 09–007. Cambridge, Mass.: Industrial Performance Center, Massachusetts Institute of Technology. November. Retrieved from http://web.mit.edu/nse/lester/media/EIP_09-007.pdf (Accessed March 3, 2014).

Lester, R.K., & Hart, D.M. (2011). *Unlocking Energy Innovation: How America Can Build a Low-Cost, Low-Carbon, Energy System*. Cambridge, Mass.: MIT Press.

Lim S.S., Vos, T., Flaxman, A.D., Danaei, G., Shibuya, K., Adair-Rohani H., Ezzati, M., et al. (2012). "A comparative risk assessment of burden of disease and injury attributable to 67 risk factors and risk factor clusters in 21 regions, 1990–2010: A systematic analysis for the Global Burden of Disease Study 2010." *The Lancet*, 380:2224–51.

Lundvall, B.A. (1992). *National Systems of Innovation. Towards a Theory of Interactive Learning*. London: Pinter.

Lutzenhiser, L. (2009). *Behavioral Assumptions Underlying California Residential Sector Energy Efficiency Programs*. Oakland, Calif.: California Institute for Energy and Environment.

Margolis, R.M., & Kammen, D.M. (1999). "Evidence of under-investment in energy R&D in the United States and the impact of federal policy." *Energy Policy*, 27:575–84.

McKeown, R. (2007). "Energy myth two – the public is well informed about energy." In *Energy and American society – Thirteen myths*, edited by B.K. Sovacool & M.A. Brown, pp. 51–57. Dordrecht, The Netherlands: Springer Science+Business Media.

Metz, B., Davidson, O.R., Bosch, P.R., Dave, R., & Meyer, R.A., eds. (2007). *Climate Change 2007: Mitigation. Contribution of Working Group III to the Fourth Assessment Report of the Intergovernmental Panel on Climate Change*. Cambridge: Cambridge University Press.

MIT. (2011). *The Future of Natural Gas. An Interdisciplinary MIT Study*. Cambridge, Mass.: Massachusetts Institute of Technology.

Narayanamurti, V., Anadon, L.D., Breetz, H., Bunn, M., Lee, H., & Mielke, E. (2011). *Transforming the Energy Economy: Options for Accelerating the Commercialization of Advanced Energy Technologies*, Energy Technology Innovation Policy Group. Cambridge, Mass.: Belfer Center for Science and International Affairs, John F. Kennedy School of Government, Harvard University.

Narayanamurti, V., Anadon, L.D., & Sagar, A. (2009). "Transforming energy Innovation." *Issues in Science and Technology*, Fall: 57–64.

Narayanamurti,V., Odumosu, T., & Vinsel, L. (2013). "The discovery-invention cycle: Bridging the basic/applied dichotomy." Discussion Paper 2013–02, Science, Technology, and Public Policy Program. Cambridge, Mass.: Belfer Center for Science and International Affairs, John F. Kennedy School of Government, Harvard University, February 2013.

NAS. (2009). *America's Energy Future: Technology and Transformation*, Washington, D.C.: National Academies Press.

Nelson, R., & Rosenberg, N. (1993). *National Innovation Ecosystems: A Comparative Analysis*. Oxford: Oxford University Press.

Nemet, G.F. (2009). Interim monitoring of cost dynamics for publicly supported energy technologies. *Energy Policy*, 37(3):825–35.

New York Times, July 13, 2011, Wald M.L. & Broder, J.M. "Utility shelves ambitious plan to limit carbon." Retrieved from http://www.nytimes.com/2011/07/14/business/energy-environment/utility-shelves-plan-to-capture-carbon-dioxide.html?_r=0 (Accessed November 1, 2013).

NSF. (2008a). "Research and development: National trends and international linkages." *In National R&D Trends. Science and Engineering Indicators 2008*. National Science Foundation. Retrieved from http://www.nsf.gov/statistics/seind08/c4/c4s1.htm (Accessed March 3 2014).

NSF. (2008b). "Research and development: National trends and international linkages. business R&D." In *Science and Engineering Indicators 2008*. National Science Foundation. Retrieved from http://www.nsf.gov/statistics/seind08/c4/c4s3.htm (Accessed March 3, 2014).

NSF. (2010). *Business Research and Development and Innovation Survey (BRDIS)*. National Science Foundation. Retrieved from http://www.nsf.gov/statistics/srvyindustry/ (Accessed March 3, 2014).

OECD. (2008). *Nuclear Energy Outlook*. Paris: Organisation for Economic Co-operation and Development.

Ogden, P., Podesta, J., & Deutch, J. (2008). "A new strategy to spur energy innovation." *Issues in Science and Technology*, Winter. Retrieved from http://www.issues.org/24.2/ogden.html (Accessed November 1, 2013).

Pacala S., & Socolow R. (2004, August 13). "Stabilization wedges: Solving the climate problem for the next 50 years with current technologies." *Science*, 305:968–72.

PCAST. (2010). *Report to the President on Accelerating the Pace of Change in Energy Technologies Through an Integrated Federal Energy Policy*, President's Council of Advisors on Science and Technology. Executive Office of the President.

Rehfuess, E., Corvalan, C., & Neira, M. (2006). "Indoor air pollution: 4,000 deaths a day must no longer be ignored." *Bulletin of the World Health Organization*, 85(7): 508–9.

Simeone, N. (2009). *Defense Department Reduces Dependence on Fossil Fuels*. U.S. Department of Defense, American Forces Press Service, April 15, 2009.

Stavins, R. (2010). "Both are necessary, but neither is sufficient: Carbon-pricing and technology R&D initiatives in a meaningful national climate policy." Retrieved from http://belfercenter.ksg.harvard.edu/analysis/stavins/?p=827 (Accessed March 3, 2014).

Tversky, A., & Kahneman, D.K. (1974). "Judgment under uncertainty: Heuristics and biases." *Science*, New Series, 185(4157):1124–31.

UNDP. (2005). "Energizing the Millennium Development Goals: A guide to energy's role in reducing poverty." New York: United Nations Development Programme.

UNEP SEFI & Bloomberg New Energy Finance. (2010). *Global Trends in Sustainable Energy Investment 2010 Report*. New York: United Nations Environment Programme and Bloomberg New Energy Finance.

Unruh, G.C. (2000). "Understanding carbon lock-in." *Energy Policy*, 28:817–30.

U.S. Census Bureau. (2010). *U.S. International Trade Data*. Retrieved from http://www.census.gov/foreign-trade/data/index.html (Accessed March 3, 2014).

Velikhov, E.P., Gagarinksi, A.Y., Subbotin, S.A., & Tsibulski, V.F. (2006). *Russia in the World Energy of the XXI Century*. Moscow, Russia: Izdat Nuclear Science and Engineering Publishers.

Weiss, C., & Bonvillian, W.B. (2009). *Structuring an Energy Technology Revolution*. Cambridge, Mass.: MIT Press.

Wheeler, T., & von Braun, J. (2013). "Climate change impacts on global food security." *Science*, 341:508–13.

Wong, E. (2013). "Cost of environmental damage in China growing rapidly amid industrialization." *The New York Times*, March 29, 2013.

World Bank. (2012). *Turn Down the Heat: Why a 4°C Warmer World Must be Avoided.* Washington, D.C.: World Bank.

World Bank. (2013). *Turn Down the Heat: Climate Extremes, Regional Impacts, and the Case for Resilience*. Washington, D.C.: World Bank.

2

Expanding and Better Targeting U.S. Investment in Energy Innovation: An Analytical Approach

LAURA DIAZ ANADON, GABRIEL CHAN, AND AUDREY LEE

In Chapter 1 we describe the urgent challenges posed by the current U.S. energy system and make the case that accelerated innovation in energy technologies will be essential in meeting these challenges. In this chapter, we explore the question of determining the appropriate role for the U.S. government in fostering energy innovation.[1] This is a complex issue because the U.S. energy system encompasses many functions (e.g., invention and discovery, R&D spillovers, economies of scale, clustering, learning by doing, deployment, etc.); government actions to facilitate innovation interact with private actors in many markets; there are a multiplicity of actors (e.g., national and regional public organizations, private firms, universities, consumers); there is deep uncertainty in future technology performance rooted in many factors (e.g., the rate of technological change, economic growth, energy prices, consumer adoption, etc.).

In this book we take a systems approach to the innovation process. The chapter starts by focusing on selected key government functions, and in Chapter 6, we integrate the analysis from this chapter with analysis and holistic recommendations that cut across the entire system. Our analysis in this chapter starts with the energy research, development, and demonstration (RD&D) investments of the U.S. Department of Energy (DOE) – the single largest public funder of energy RD&D in the United States. Despite the broader scope of DOE's jurisdiction, which we trace to its historical roots, we limit our analysis to only the department's energy RD&D funding activities. After describing DOE's current role investing in energy RD&D, we analyze how DOE makes decisions about its energy RD&D investments. We argue that a method to inform public energy RD&D funding decisions that allows for an integrated, consistent, and transparent analysis among competing technology areas is needed. We present and implement such a method, relying on a broad set of detailed expert elicitations and a comparably detailed energy-economic model of the

[1] This question has been the subject of several notable past studies (PCAST, 1997, 2010; American Energy Innovation Council, 2010; Weiss & Bonvillian, 2009; Lester & Hart, 2011).

U.S. economy. The method we propose in this chapter has broader applicability and could be used in similar RD&D funding institutions in the United States and abroad.

2.1. Investments in Energy RD&D at the U.S. Department of Energy: Status Quo, History, and Decision Making

2.1.1. Status Quo

Despite the well-established benefits of advanced energy technologies, for the majority of the last 30 years the U.S. government has made limited investments in energy RD&D. From 1986 to 2006, DOE's annual RD&D spending was one-third of its annual spending in the wake of the 1973 global oil crisis during (Figure 2.1).[2] Between 2008 and 2013, annual funding has been slowly increasing to around $3 billion a year (or $4.7 billion if investments in the Basic Energy Science program are included). A notable break in the trend of DOE funding was the American Recovery and Reinvestment Act (ARRA) of 2009, which included a one-time package of energy RD&D investments comparable in scale to the annual energy RD&D of the late 1970s.

The combined investments of the other seven U.S. federal agencies involved in energy RD&D and "basic science" was $1.2 billion in 2012. Of this, $940 million was funded through the Department of Defense (DOD) (Energy Innovation Tracker, 2013), highlighting the important role of the DOD in shaping the pace and direction of U.S. energy innovation. Different state governments also support energy RD&D, but these investments are more atomized, making them more difficult to quantify. Little information is readily accessible on aggregate state-level energy RD&D funding beyond a rough estimate of a combined $3 billion investment in 2008.

2.1.2. History

The Department of Energy was created in 1978 in response to the 1973 oil crisis as a way to consolidate energy innovation funding strategically. Prior to 1978, energy innovation funding was funded through two separate agencies, the Atomic Energy Commission (AEC) and the Environmental Protection Agency (EPA). DOE inherited the nuclear weapons program from the AEC (which had previously been developed under the Manhattan Project) and continued to fund high energy and nuclear physics research. In fact, most of DOE's spending is not in the area of energy innovation: out of the $26.3 billion appropriated to DOE in the 2012 budget, less than one-fifth ($4.7 billion) was dedicated to energy RD&D and Basic Energy Sciences; $16.8 million supported atomic energy defense activities. It is not surprising then that energy innovation has not historically been a major area of focus for DOE's leadership and, therefore, there has been limited focus on planning how the agency can best

[2] Unless noted otherwise, all $ will be in constant 2010$.

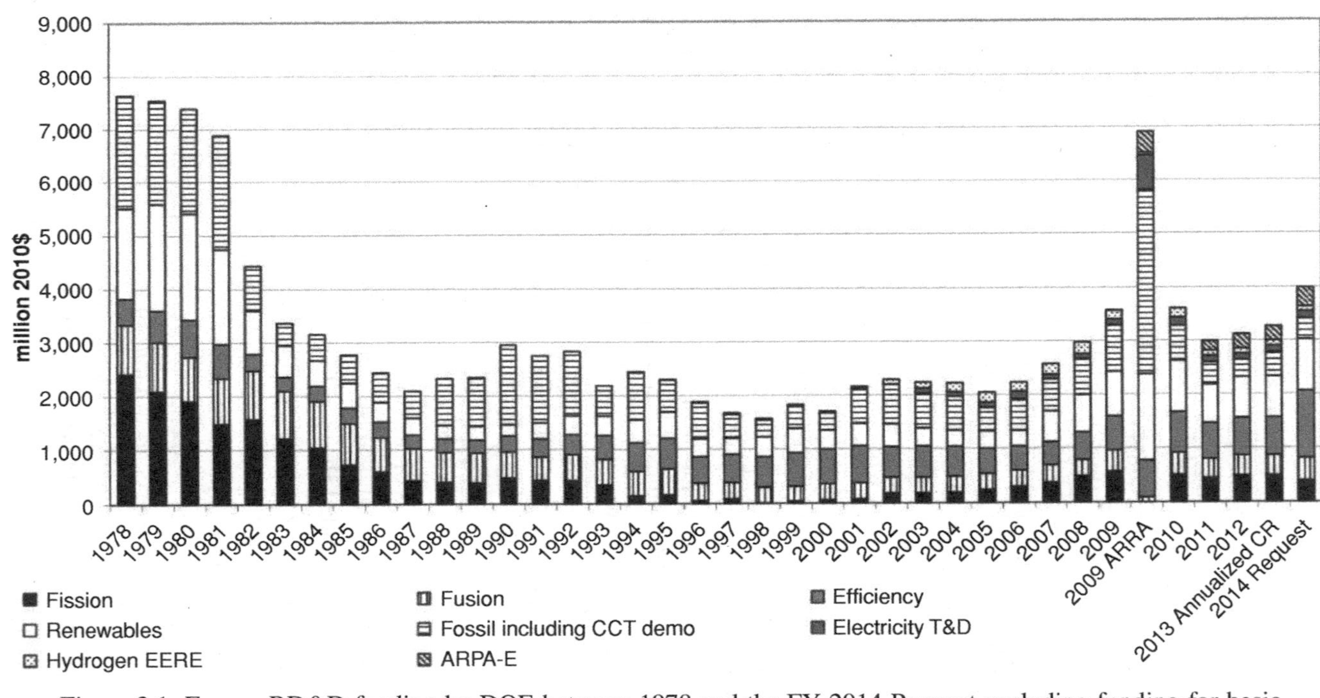

Figure 2.1. Energy RD&D funding by DOE between 1978 and the FY 2014 Request excluding funding for basic energy sciences. CR = continuing resolution. (Gallagher & Anadon, 2013).

fulfill this part of its mission. It is also well known that different administrations have had different ideological beliefs about both the role of the government in fostering energy innovation and on which technologies would be the most promising investments. Most notably, in the early 1980s, the Reagan administration reduced its energy RD&D investments drastically. Changes in administration ideologies have been major contributors to the volatility of total investments and their allocation to different technologies over time. Across 13 technology areas (coal, carbon sequestration, gas, petroleum, transportation, buildings, industry, fission, fusion, solar, wind, geothermal, and bioenergy), average year-on-year percent change in RD&D funding between 1978 and 2013 was 4% with a standard deviation of 37%. Over this time period, funding in these technology areas changed from year to year by at least 29% in one-in-five cases and by more than 47% in one-in-ten cases. The largest variance in year-on-year percent changes in funding have been in geothermal, nuclear fission, and nuclear fusion technologies.[3]

Volatility in investments is not necessarily a bad feature if volatility indicates that decision-making processes are responsive to new information or economic circumstances. However, in the case of DOE's RD&D investments, the observed volatility in year-on-year RD&D investments has likely lowered the overall effectiveness of these investments. Because of the long-term uncertain nature of RD&D, volatility in funding can cut programs short that might otherwise succeed if given more time. This point may be particularly important for understanding the effectiveness of the relatively large one-time investments of the stimulus package. Current annual budgets have not sustained the funding provided by ARRA beyond the policy's cutoff date for implementing funded projects, which will likely diminish this investment's effectiveness. Figure 2.2 shows the variation in the 13 technology areas investigated.

Even though DOE was created in 1978, U.S. government support for energy RD&D started a few decades earlier. Between 1945 and 1961 energy RD&D investments were largely limited to nuclear energy. The 1960s saw the birth of a modest coal RD&D program in the Office of Coal Research. The 1940s and the 1950s also saw many of the national laboratories that were created during WWII expand their mission to include energy RD&D (see Chapter 3 for an analysis of the role of DOE national laboratories).[4] In general, between the 1940s and the early 1970s there was a perception that the United States had cheap and abundant energy supplies, and that federal government support for basic science through the National Science Foundation combined with support for nuclear RD&D through the Atomic Energy Commission

[3] The variance in year-on-year funding is even larger in hydroelectric technologies, but we have excluded it from this analysis as it has historically been funded at a lower level than the other 13 technology areas described previously. We have also excluded ARRA funding from this analysis as this was a one-time set of investments that did not fall into a single budget cycle, occur in a single fiscal year, or originate from the typical DOE budget decision-making process.

[4] Two more laboratories were added in 1977 (the National Renewable Energy Laboratory, focused on renewable energy) and in 1984 (the National Energy Technology Laboratory, focused on fossil technologies).

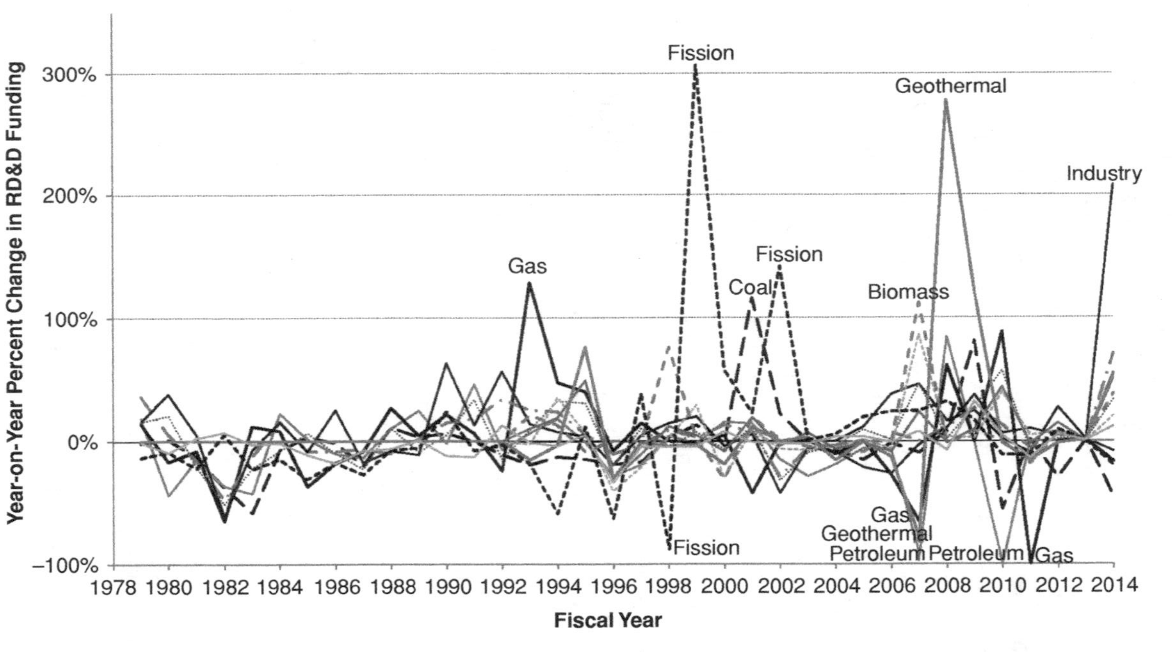

Figure 2.2. Year-to-year volatility of DOE support for RD&D in nine technology areas between 1978 and the 2014 budget request. The vertical axis has been truncated although the values are included in the calculations.

would be sufficient to meet the nation's long-term energy needs (Dooley, 2008). This perception changed with the Arab Oil Embargo of 1973. Prodded by national security concerns and the realization of the United States' vulnerability to future oil supply disruptions, the United States began a reevaluation of its energy technology program and the role that RD&D could play in national energy policy. Following 1973, federally funded energy RD&D increased significantly and became more diversified. The first federal solar programs for terrestrial applications, geothermal programs, and energy conservation programs were started in 1974 (Stewart et al., 1983). During this period, the U.S. government's role in innovation was expanded beyond basic research for the first time to also include applied research and the development of small, one-of-a-kind pilot facilities (Dooley, 2008). From the 1970s through the late 1990s, owing in part to the increasing cost of nuclear energy and the highly visible Three Mile Island accident of 1979, public investment in nuclear fission RD&D decreased steadily. Between 1985 and the proposed 2014 budget,[5] 20% of cumulative energy RD&D funding has been allocated in end-use efficiency in transportation, industry, and buildings; 28% in fossil energy (coal, petroleum, and natural gas) and carbon capture; 15% in nuclear fusion; 12% in nuclear fission; 19% in renewables; 2% in hydrogen; 2% in electricity transmission and distribution; and 2% in the Advanced Research Projects Agency-Energy (ARPA-E) (Gallagher & Anadon, 2013).

2.1.3. Making Decisions about Investments in Energy RD&D at DOE

Federal funding is an important mechanism to support high-risk RD&D projects with social benefits that would not otherwise attract private investment. Several prominent studies over the past couple of decades have called for greater U.S. government spending on energy technology RD&D. Some of these studies have also considered portfolios of energy RD&D investments to reduce future greenhouse gas (GHG) emissions (PCAST, 1997, 2010; NCEP, 2004, 2007; APS, 2008; American Energy Innovation Council, 2010). Although all of the studies cited conclude that directing additional resources toward accelerating energy technology innovation are warranted, none of them offer robust analytic support for their recommendations. Nor do these studies provide quantitative assessment of the potential outcomes of additional government RD&D investment, making them of minimal use to any skeptical decision makers setting RD&D investment levels and allocations. In other words, none of these studies discuss how to allocate a portfolio of RD&D investments based on an evaluation of prospective benefits with inherent uncertainty, nor do these studies consider how the future benefits of RD&D investments depend on the interplay of technologies in the marketplace.

The decision processes that have resulted in the annual DOE energy RD&D investment portfolios presented in Figures 2.1 and 2.2 have changed significantly over time.

[5] These figures include the American Recovery and Reinvestment Act of 2009 funding.

We argue that the future development of decision making at DOE would benefit from the insights of an integrated, consistent, and transparent decision framework.

Before the Bush administration, annual DOE RD&D investment decisions were made in a three-step process. In the first step, DOE program offices (Energy Efficiency & Renewable Energy, Nuclear Energy, Fossil Energy, and Electricity Delivery & Energy Reliability) self-assessed estimates of the benefits from investing in their own RD&D programs. These estimates were developed in independent frameworks in each of the program offices, relying on different sets of assumptions (e.g., some programs used estimates of much higher oil prices than others). This first step was followed by action under the direction of the Secretary of Energy, which collated the different estimates and submitted them to the Office of Management and Budget (OMB), the Office of Science and Technology Policy (OSTP), and to Congress, accompanying the President's annual budget request. The third step of this process was outside of DOE's control: Congress modified funding for different programs in its process of approving the budget for the next year. These modifications were made on the basis of individual Congressional members' and political parties' interests and priorities.

The second and third steps of the budget process have not changed over time: DOE submits an annual budget justification accompanying the president's budget request, and Congress modifies the request. It is the first phase of the process that has changed over time and where we believe the approach that we present could most improve the process. From the perspective of individual program offices, due to the real and perceived competition for funds between offices, cost reduction estimates from other offices were seen as purposefully overly optimistic and designed to attract additional investment. In the early 2000s, the Bush administration created an effort to standardize economic assumptions underlying estimates of future benefits of technology programs and begin understanding interactions between programs. At about the same time, a separate effort at DOE headquarters was designed by the Climate Change Technology Program. This effort relied on a different model than the program offices typically used (the Global Change Assessment Model [GCAM]) rather than more technologically detailed models such as the MARKet ALlocation Model [MARKAL] or the National Energy Modeling System [NEMS], which do not include a climate model) and on a different set of technology assumptions that were not made available for public scrutiny. Both of these efforts to integrate benefit assessment across all of the DOE program offices were eventually discontinued. In 2007 and 2008 the Office of Energy Efficiency and Renewable Energy (EERE), the Office of Fossil Energy (FE), and the Office of Nuclear Energy (NE) jointly participated in an integrated exercise to provide internal estimates on expected technology costs under different RD&D levels to modelers that would allow for a more consistent comparison of integrated benefits. But concerns about the credibility of technology cost estimates and difficulties reaching consensus ended the effort. Between 2009 and 2010 EERE continued with an integrated analysis of only its own technologies, but in 2011 the effort was cancelled, partly due to changes in the office's leadership and its priorities.

Today, DOE essentially does no systematic and consistent comparison of the expected benefits of its technology RD&D programs.

Even during the two-year period (2007–2008) when three major DOE Offices collaborated to integrate the expected benefits of their programs, DOE did not consider the uncertainty associated with the outcomes of their different applied RD&D programs, even though EERE supported the development of the SEDS energy system model to capture uncertainty. More importantly, DOE did not use a consistent method to estimate the impact of RD&D on future costs. Some of the challenges of these early efforts could have been addressed if the process had been made more transparent and independent.

In part to address some of these concerns, a 2007 report of the National Research Council highlighted that the applied RD&D programs of DOE should consider the societal impacts of technology innovation resulting from government energy RD&D as well as the probability of such impacts (NRC, 2007).

In the rest of this chapter we present and implement a methodology for estimating the benefits of investing in a range of energy technologies. We argue that the use of and insights resulting from this method can be used to help better understand the robustness of different energy RD&D portfolios under different assumptions about uncertainty and future policies. Further, we make the case that this type of analysis should become institutionalized at DOE as it can feasibly be implemented in a transparent and independent manner.

The methodology we propose consists of (1) designing and conducting expert elicitations to obtain estimates of the impact of public RD&D on the possible future costs and performance of seven classes of energy technologies; (2) using these expert elicitation estimates in an energy-economic model to estimate the benefits of various RD&D portfolios; and (3) calculating the optimal allocation of investments across technology areas that would be needed to increase the probability of achieving certain targets (reductions of CO_2 emissions and oil imports, among others). The methodology takes into account uncertainty in technological progress, interactions between technologies in the marketplace, and concerns for transparency and consistency in the estimates.

2.2. RD&D Investments by the Department of Energy Cannot Be Evaluated in Isolation

Over the past 50 years, publicly funded basic and applied research has played an important role in the growth of the agriculture, chemicals, life sciences, and information technology industrial sectors in the United States (Mowery, Nelson, & Martin, 2010; Henderson & Newell, 2011). Retrospective analysis of publicly funded energy research and development programs has found a right-skewed distribution in the returns to RD&D investments, leading to the finding that a relatively few number of successful programs bring about benefits that outweigh the costs of the entire portfolio. Market failure that stems from two sources of inappropriable benefits leads the private sector to systematically under-invest in developing new energy technologies: (1) in the absence of effective government policies, the environmental benefits of new technologies do not accrue to private investors, and (2) private innovators develop intangible knowledge that has spillover benefits for other innovators that the investing innovators are not able to appropriate.

RD&D policy alone would not be a cost-effective policy strategy for reducing GHG emissions to levels proposed in recent policy goals. Further, RD&D policies can be made more effective in increasing the rate of advancement of new technologies and ideas with complementary market pull policies (Mowery & Rosenberg, 1979; Goulder & Schneider, 1999; Holdren & Baldwin, 2001; Fischer & Newell, 2008; Anadon & Holdren, 2009). Before achieving widespread diffusion, novel energy technologies are often faced with barriers that can include costs above those of incumbent technologies; lack of supporting infrastructure; slow capital stock turnover of incumbent technologies; lack of information about new technologies for consumers; and financing constraints. Deployment policies, including market-based environmental regulation, performance or technology standards, and tax policies, can reduce the barriers to deployment by lowering the uncertainty in the demand for new technologies. On the other hand, unstable or unpredictable deployment policies can discourage investment in innovation and counteract RD&D efforts. Therefore, an analysis of the expected benefits of energy RD&D investments cannot be complete without an analysis of complementary deployment policies. These benefits will also depend on other market conditions, for example, oil and gas prices. The methodology of estimating the benefits of energy RD&D that we introduce below explores these dependencies through scenarios of policy options and market conditions within the framework of an energy system model (MARKet ALlocation [MARKAL]).

In this work, however, we do not incorporate the effect that "demand pull" policies have on stimulating RD&D in the private sector. Environmental policies, such as putting a price on carbon emissions, stimulate additional private investment in finding cost-effective ways to reduce environmental externalities. The private-sector energy innovators we surveyed reported that expected prices were by far the largest drivers of their investment decisions (see Chapter 4 for more detail). However, there is insufficient empirical evidence to determine how large of an increase in private investment would result from policies designed to complement technology policies, and how much this feedback would affect the future cost and performance of energy technologies. Because our modeling does not include this feedback, our analysis likely underestimates the benefits of demand-side policies, and likely overstates the costs of reaching particular climate targets in the future. The new market opportunities afforded by policies that do not prescribe technology choices are likely to stimulate innovation in ways that are unpredictable. In our modeling, we consider instead how these "demand pull" policies affect the diffusion of technologies through their impact on relative prices of technologies rather than through progress in technology prices themselves.

2.3. Review of Methods to Evaluate Budgets for Energy Technology Innovation

In this section, we review the approaches that have been proposed in the literature to aid decisions about investments in energy innovation. This sets up the discussion in the following section of the methodology we developed of eliciting technology experts for RD&D budgets and outcomes, and evaluating those estimates with a model of the U.S. energy system.

A large literature of empirical studies has found that technology R&D investments yield large returns to society. Griliches' review of various studies estimating the return to R&D investments found that there is a 10% to 80% *social* return on investment for technology R&D (Griliches, 1992). In the context of DOE's energy RD&D investments, one study found that the benefits of just six of DOE's energy efficiency programs from 1978 to 2000 produced economic benefits almost four times larger than the cost to the government of supporting all efficiency research programs during the time period considered (NRC, 2001). Despite finding persistently positive social returns to R&D, the literature on retrospective studies of R&D programs finds a wide range of benefits, reflecting the inherent uncertain nature of innovation. Three studies from the literature of particular relevance have tried to estimate the value of supporting energy technology R&D prospectively (or into the future).

Schock et al. (1999) calculated the value of low-carbon R&D as the difference between the cost of limiting CO_2 emissions to different stabilization levels[6] with and without advanced technologies, assuming a "probability of R&D success" ranging from 0.1 to 1. The potential value of U.S. low-carbon energy R&D to reduce climate change mitigation costs was estimated to be \$6.4 to \$9 billion annually. The paper also estimated the value of R&D to reduce the costs from oil price shocks and urban air pollution. Nemet and Kammen used the results from 550 ppmv CO_2 stabilization scenario of Schock et al. (1999) to recommend that \$6.7 to \$30.1 billion be invested in low-carbon technology R&D (Kammen & Nemet, 2005a, 2005b). An important weakness of this approach is that it is based on speculative assumptions about the likely rate of R&D success, and does not differentiate among technologies in this respect, making it impossible to extract conclusions about budget allocations by technology.

Davis and Owens (2003) used the concept of real options to estimate the option value of investments in renewable energy R&D. They considered the volatility of natural gas prices (and its impact on the price of natural gas power) and made assumptions about the effect of R&D on reducing the uncertainty around achieving particular technology improvements. The result of this estimate was that U.S. renewable energy R&D would be worth \$5.4 billion. The authors then introduced assumptions about how the rate of renewable electricity cost reductions would increase (from 1% per year to 4% per year) to estimate the optimum yearly level of renewable energy R&D

[6] No limit with a 35% probability; 450 parts per million of CO_2 by volume (ppmv) with a 5% probability; 550 ppmv with a 25% probability; 650 ppmv with a 20% probability; and 750 ppmv with a 15% probability.

to be $1.4 billion. This approach is similarly based on speculative assumptions about the degree of progress that will result from increased R&D, and does not differentiate among different technologies.

Finally, Blanford (2009) estimates the optimal allocation of R&D funds for renewable energy, nuclear energy, and coal with carbon capture and sequestration (CCS) by defining two states for the cost of those technologies ("optimistic" and "pessimistic"), assuming several different functional forms for the probability of achieving the "optimistic" cost of technology, and maximizing the expected value of global GDP in a general equilibrium model (MERGE [*m*odel for *e*stimating the *r*egional and *g*lobal *e*ffects of greenhouse gas reductions]).

Although this approach differentiates among three broad classes of technology, there is the major simplifying assumption of only two possible future costs for these technologies, and the result is sensitive to the assumptions about how R&D investments would affect the probability of those two future states.

In short, other analyses do not claim to present a transparent and consistent methodology for linking R&D investments and their uncertain technology outcomes. Others have focused on forecasting the impact of R&D investments on future cost(s) and performance of a technology. To the best of our knowledge, four methods have been used: (1) expert elicitation, which can be used to predict potential outcomes of hypothetical RD&D programs, thus lending insight into uncertainty (Morgan & Henrion, 1990); (2) decomposing long-term trends in technology cost (Nemet, 2006; McNerney, Trancik, & Farmer, 2011) and using them to infer future technology developments; (3) monitoring "precursors," such as scientific publication and patent data, that strengthen or weaken a hypothesis that a technological change will take place (Martino, 1987); and (4) technology roadmapping, in which experts lay out particular technical barriers to be overcome and attempt to estimate the costs of programs to overcome those barriers (Garcia & Bray, 1997).

Several studies have elicited expert opinions about the relationship between different levels of U.S. energy R&D spending and the probability of achieving set cost and performance benchmarks for carbon capture and storage (Baker, Chon & Keisler, 2009b), photovoltaics (Curtright, Morgan & Keith, 2008; Baker, Chon & Keisler, 2009a), biofuels (Baker & Keisler, 2009), nuclear energy (Baker, Chon, & Keisler, 2008), and vehicle battery technologies (Baker, Chon, & Keisler, 2010). These studies evaluate low, medium, and high levels of funding, as defined by the consulted experts (Baker, Chon, & Keisler, 2008; Baker, Chon, & Keisler, 2009a; Baker & Keisler, 2009; Baker, Chon, & Keisler, 2010), and 2- and 10-fold increases over current spending (Curtright, Morgan, & Keith, 2008). Experts could also lend insight into other factors affecting innovation such as spillovers from other areas (see Clarke, Weyant, & Birky, 2006) for a discussion of spillovers), and the socio-technical dynamics of innovation (e.g., networks, community building) (see Geels, Hekkert, & Jacobsson, 2008 for a discussion of the importance of socio-technical factors).

Previous studies have combined expert elicitations about the relationship between R&D investments and technical change with integrated assessment models to estimate the marginal cost of abating CO_2 emissions under different scenarios of R&D investment in individual technologies (e.g., Baker, Chon, & Keisler, 2009a, 2009b study R&D in advanced solar and carbon capture and storage using the GCAM model, previously MiniCAM). Another group is conducting expert elicitations to account for R&D-induced technical uncertainty using the World Induced Technical Change Hybrid (WITCH) integrated assessment model (Bosetti et al., 2009). Finally, researchers at the Climate Change Technology Program at the U.S. Department of Energy and the Pacific Northwest National Laboratory presented two approaches to assess the tradeoffs in constructing a portfolio of technology acceleration goals to guide DOE's budgeting process. In one approach, the value of a technology was defined in terms of the quantity of reduced carbon emissions to which its use can be attributed. In the other, the value of a technology was defined as the cost reduction in meeting a fixed emission reduction target. In both approaches the value of technology was quantified by an internal elicitation of DOE employees and model runs from GCAM. The portfolio was built by rank-ordering return on investment to individual technology R&D programs (Pugh et al., 2010).

The unique contributions of this work (described in Section 2.4) to existing methodologies allow us to assess energy RD&D allocations to different energy RD&D stages and technology areas. We account for the uncertainty around the benefits of investing in a portfolio of technologies and examine a wide range of budget, deployment, and market scenarios.

2.4. Method for Providing Insights about the Returns to and Allocations of Federal Energy RD&D Investments

The U.S. DOE could implement a more consistent, transparent, and comprehensive method to inform its energy RD&D budget proposals. We now present a multistep method we developed to support energy RD&D decisions, and then show the results of its implementation over seven technology areas in Section 6.

2.4.1. Step 1: Designing and Fielding Expert Elicitations

We designed and conducted expert elicitations (or formal surveys) with more than 100 experts[7] on seven technology areas: fuels and electricity from biomass (bioenergy, for short); different types of utility scale energy storage (or storage); residential, commercial, and utility scale photovoltaic technologies (solar); efficiency in commercial buildings (buildings); nuclear power from Generation III/III+, Generation IV,

[7] 100 experts gave total and partial answers to the different surveys: 30 participated in the nuclear survey, 25 in the storage survey, 12 in the bioenergy survey, 9 in the buildings survey, 9 in the vehicles survey, 12 in the fossil energy survey, and 11 in the PV survey.

and small and medium reactors (nuclear); coal and natural gas electricity production with and without carbon capture and storage (fossil); and vehicle technologies (vehicles). We invited leading researchers and practitioners in each of the seven fields from academia, the private sector – including small and large firms – and the U.S. national laboratories to give us their input. We asked these experts to estimate technical cost and performance metrics in 2030 under a business as usual (BAU) scenario for federal RD&D funding for that technology, to recommend the level of annual RD&D funding and allocation that would be necessary to increase the commercial viability of the technologies in question, and to revise their 2030 technology projections under different hypothetical budgets.

We were unable to include other energy technologies in DOE's investment portfolio, notably wind power and geothermal power, in this study owing to limited resources. The seven technology classes listed above were given priority, because wind power already has substantial commercial deployment, and a 2008 report by the National Renewable Energy Laboratory (NREL) highlighted research priorities and technology potential of wind power (DOE, 2008). Similarly, a comprehensive report provided recent insight into the future of geothermal power (MIT, 2006).

Our buildings elicitation asked experts about RD&D investments in building technologies for both commercial and residential buildings, and for the necessary RD&D investments on different parts of the building envelope, lighting technology, energy metering programs, building modeling studies, and so forth. Our questions about technology outcomes, however, were limited to the cost of reducing the energy demand for heating and cooling (or HVAC [heating, ventilation, and air conditioning], which represents 39% of energy use in residential buildings and 32% of energy use in commercial buildings). Even though lighting is the second largest energy user in both types of buildings, the improvements of lighting technologies resulting from the budgets recommended by the experts were not included in the analysis. In addition, the experts consulted decided to provide their insight about the role of RD&D on the cost of commercial and not residential buildings (commercial buildings used 18% of primary energy in the United States in 2006, while residential buildings used 21% [EIA, 2010e]). Therefore, the benefits from the RD&D investments in buildings are lower than they would otherwise be in a more holistic analysis of building technologies.

In addition, while our elicitations did ask about basic research needs for specific technology areas, the recommendations were not meant to incorporate all needs in the Basic Energy Science (BES) program. Thus our analysis does not include insights on overall funding for the Office of Science or the Basic Energy Sciences program.

2.4.2. Step 2: Qualitative Reviews of Elicitation Results

We asked an additional set of experts (a total of 23), with experience thinking about investments on a range of technology projects, for their feedback on the results from each of the seven expert elicitations. Their affiliations included DOE programs, venture capital firms, and U.S. Congressional committees. These qualitative reviews helped us interpret the elicitation results and served to expand expert input in our work.

2.4.3. Step 3: Selection of Three Types of Experts for Each Technology Area

We selected the answers from three experts for each technology area, representing optimistic, "middle-of-the-road," and pessimistic forecasts (encompassing the full range of answers to the extent practical). We used this smaller set of estimates to create three representative technology profiles in the Brookhaven National Laboratory 10-region MARKAL model. We chose this model because it is used by the U.S. DOE for energy policy and Government Performance and Results Act analyses. In addition to defining these three types of technology profiles, we also developed four different RD&D funding scenarios, defined by the elicitation questions. Each combination of technology profile and budget scenario has different technology cost and performance characteristics, as specified by the experts consulted. We also designed a sampling algorithm (see Section 8 for more information) to allow us to estimate the uncertainty around the benefits of a portfolio of investments, based on the experts' responses. Using MARKAL, we calculated the following benefits of budget scenarios beyond BAU funding, accounting for uncertainty: reductions in CO_2 emissions, reductions in CO_2 prices, reductions in Clean Energy Standard (CES) credit prices, reductions in oil imports, and increases in consumer surplus.

2.4.4. Step 4: Defining Policy and Market Scenarios

We defined policy and market scenarios to examine how the benefits of the RD&D programs would change under different non-RD&D policy scenarios. Particular policies evaluated include limits on CO_2 emissions, a clean energy standard, more stringent building codes, and more stringent transportation fuel economy standards. The market scenarios we studied varied the prices of oil and natural gas; we included one scenario with high oil and gas prices and one scenario with high oil and low gas prices.

2.4.5. Step 5: Sampling and RD&D Portfolio Optimization

After selecting experts and scenarios, we fit formal probability distributions to the elicited statistics estimated by the experts (the 10th, 50th, and 90th percentiles of cost). We then used a sampling technique to represent uncertainty without requiring prohibitive computational capacity. Finally, we implemented an optimization scheme over the MARKAL model output to estimate the allocation of RD&D investments that would maximize different societal outcomes under different budget constraints.

2.4.6. Relationship between this Work and Previous Work

We used expert elicitation to obtain inputs about technical change resulting from government RD&D investments, and the uncertainty around these inputs, for two reasons. First, we could not rely on extrapolation of cost reductions from historical data on the evolution of a technology because we wanted to capture the possibility that technologies supported by public RD&D investment may advance through new, potentially breakthrough pathways. Second, expert elicitation allows for the quantification of uncertainty (or at least that portion of the uncertainty perceived by experts in the field) and can help decision makers collect information in cases where observable data is sparse or unreliable, and potentially useful data is unpublishable or proprietary (Chan et al., 2010; Anadon et al., 2012).

Although the work reviewed in Section 3 uses the concept of insurance and option value, these studies restricted their analyses to the early stages of innovation (i.e., research and development). The analysis presented in this work considers the federal funding necessary to support use-inspired basic research, applied research and development, pilot-scale experiments, and commercial demonstration on energy technologies, the entirety of which we refer to as "RD&D." We included use-inspired basic research and demonstration because, as discussed in Chapter 1, innovation is a complex and interconnected process, in which research and development investments are important but not sufficient by themselves. The intent of a commercial scale demonstration is to show that the technology can be effectively used at full commercial scale in an economically useful way (i.e., at reasonable cost). However, such a scale-up is also the first opportunity to learn how the technology works at full scale, and whether the design features are reasonably workable, controllable, and optimal. As such, the beginning of operations of a full-scale demonstration plant is an experiment, designed for learning and "debugging." A new plant almost always needs adjustment, particularly in control systems and their use, sometimes needs alterations in hardware, and always requires some accumulation of experience in operating the plant. Demonstration provides an opportunity to learn what changes the builders of the next plant(s) should make. There are three main differences between our approach and that of Baker et al. (2008, 2009a, 2009b), who also using expert elicitations: (1) our methodology takes into account the uncertainty around the benefits of investing in a portfolio of technologies, instead of a single technology area; (2) our elicitations asked experts to define three points of the distribution of costs in 2030, instead of the probability that particular costs would be reached, which allows us to approximate the full uncertainty around the costs; and (3) our analysis is limited to technology development, deployment, and benefits in the United States, and we make no assumptions about how technology would diffuse across borders or policies in other countries. There are also significant differences between our work and that of Pugh et al. (2010): (1) their work uses a model of the world energy sector; (2) instead of asking one group of experts about the probability of reaching an advanced technology success stage for eight technology areas, we relied on many technical experts from all sectors that differed by technology, discussed each of the expert elicitations with another set of program managers, and disclosed the raw results of the elicitations publically; and

(3) we represented aggregate benefits from portfolios, which requires estimating the correlations between the costs of different technologies.

To make final recommendations about the size and allocation of energy RD&D investments, we combine the expert views on the relationship between RD&D and technology outcomes with the environmental and economic benefits of those outcomes under different non-RD&D policies (such as a carbon policy). In doing this, we also study the impact of non-RD&D policies on the benefits of RD&D investments through their impact on relative prices and the specification of minimum performance standards. However, as mentioned earlier, the impact of non-RD&D policies on the evolution of technology cost and performance (as hypothesized by Hicks (1932) over time was not considered. There was insufficient information from the literature to estimate how cost reductions would increase beyond those driven by RD&D investments as specified by the experts. We hypothesized that technology experts would have difficulty thinking about the impact of a renewable portfolio standard (for instance) on stimulating innovation and reducing the cost of a technology. Our elicitations did, however, ask experts about their views on non-RD&D policies. Along these lines, it is important to reiterate that RD&D investments are only one of the many government actions that can catalyze innovation.

There are very large uncertainties in economic modeling extending decades into the future. Technological and social changes will surely occur in the future that cannot readily be predicted by experts today. Therefore, these uncertainties cannot be incorporated in models, including our own. In addition, integrated assessment models of future energy and economic trajectories continue to produce widely varying results depending on the assumptions made and the structures of the particular models. The MARKAL model we used, like other energy economic models, is based on a wide range of assumptions, some of which are theoretical abstractions that cannot be empirically evaluated. The modeling results we present should be considered with these caveats in mind. Our results provide a general idea of plausible outcomes rather than specific predictions; probabilities we report are the probabilities of achieving certain results given specific sets of assumptions within the confines of this particular model. The large and irresolvable uncertainty about future economic trajectories represents another key reason for our recommendation that the U.S. government invest in a broad portfolio of energy technologies to maximize the chance of developing technologies that will prove to be attractive for widespread deployment in the broadest range of future circumstances.

2.5. Energy Technology Expert Elicitations for RD&D Portfolio Analysis Support: Details, Process, and Limitations

Expert elicitation is a formal method that can be used to obtain insight into nascent technologies with uncertain cost and performance attributes. We used expert

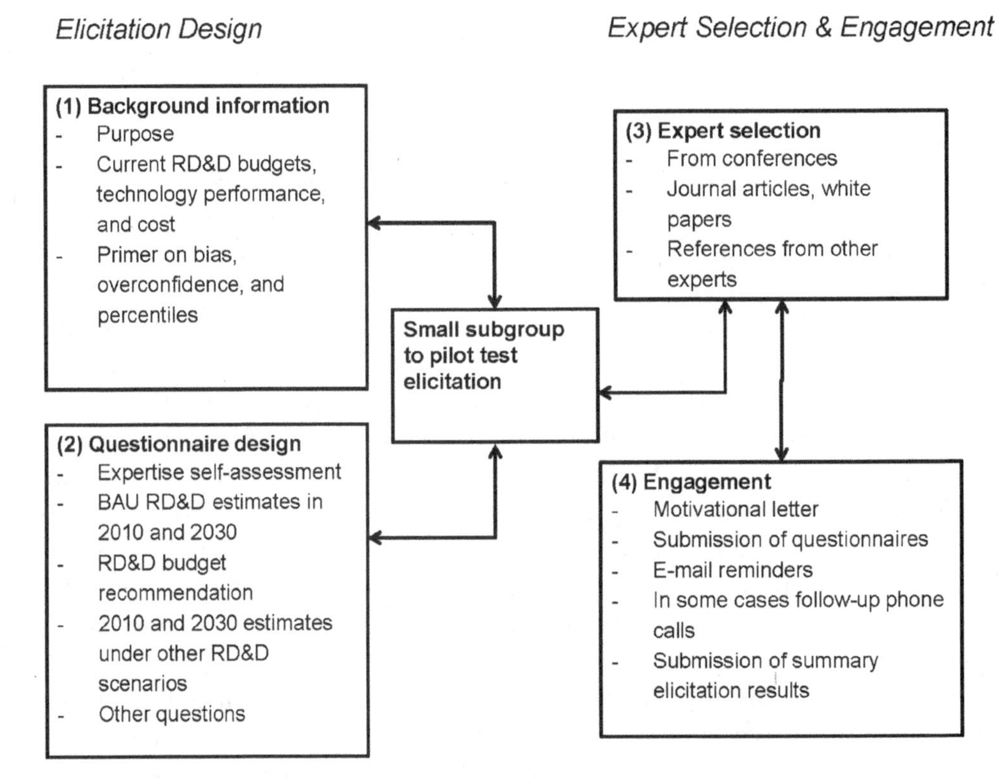

Figure 2.3. Schematic of the protocol used to design and field the expert elicitations. (Adapted from Anadon et al., 2012.)

elicitation to determine the relationship between federal energy technology RD&D funding and a range of future technologies' cost and performance attributes. Our seven elicitations were carried out using a consistent methodology, and are more comprehensive than previous studies in terms of the technology areas covered. We examined four energy supply technology areas (fossil, nuclear, solar, biofuels), two energy demand technology areas (vehicles, buildings), and one enabling energy technology area (energy storage). The elicitations were designed to encourage experts to think about technologies and research pathways broadly by leaving many questions open-ended and structuring quantitative questions in a systematic manner without over-defining specific technologies within the scope of each elicitation.

Figure 2.3 illustrates the different tasks that were involved in the design and execution of each of the seven elicitations. Designing and fielding each elicitation took between four and eight months. The first phase involved conducting research into each technology, which we summarized for expert participants in a background information section.

In the second phase, we designed the questionnaires for each technology. The questionnaire first asked experts to self-assess their expertise in a wide range of technology subfields; we used this information to explore potential biases in experts' technical assessments. The questionnaire also asked experts to recommend an aggregate level of RD&D funding, and then to estimate cost and performance metrics once in 2010 and then again in 2030 under four scenarios of RD&D funding: business-as-usual (BAU), half of the recommended budget, the recommended budget, and 10 times the recommended budget. The results of this suite of conditional cost and performance metrics allowed us to explore the sensitivity of the cost and performance estimates to different public RD&D investments. A small subgroup of experts (typically two to three) was then used to test and refine the first elicitation draft to increase our confidence that each elicitation instrument draft would be correctly interpreted by the experts (and that it would provide information in the right form to be of use in the next stages of the method). This second phase of fielding the final elicitation instrument took between two and three months.

In the third phase, we collected the names of experts from a range of sources by examining the peer-reviewed literature, national laboratory programs, university research programs, and referrals from other experts and our own program's advisory board. The participant pool we assembled covered a range of perspectives and included technical experts from the private sector, academia, and the national laboratories.

In the fourth phase, we engaged experts through email solicitations for participation. For willing participants, survey responses were conducted by mailed hard copy or through an online platform. In many cases it was necessary to send reminders and hold follow up with phone calls to clarify specific questions from the participants. Experts typically devoted between two and four hours to our study survey.

Designing and fielding an expert elicitation is a very labor-intensive process for both study designers and participating experts. Because of time and funding constraints, we were unable to cover the full range of technologies supported by DOE. Notably, we were not able to include wind power, geothermal power, concentrated solar power, advanced lighting, and industrial energy efficiency in our study. If this method were to become institutionalized at DOE to inform holistic budget decision making, it would have to expand to cover these additional technology areas. However, repeating this analysis in new technology areas would become less time-consuming with experience, as it would take significantly less time to revise the elicitation instrument than to design it and test it from scratch.

We now turn to describing in more detail the structure of the elicitation instruments. The elicitations began with a technology primer of material on current technology cost and performance, fuel costs if applicable, and a summary of current government RD&D investments in the technology area. This was done to ensure that all experts had the most recent literature fresh in their minds and to encourage them to think about the variables that we would ask them to estimate, which included costs, efficiency, and government RD&D investments and programs. The elicitation device was then divided into four or five sections of questions.

In Part 1, experts were asked to assess their expertise on specific technologies, components, and ancillary items such as feedstocks, specific technology areas, materials, products, and enabling technologies.

In Part 2, experts were asked to identify commercially viable technologies in 2010, as well as their cost and performance that would result from a continuation of 2010 federal funding and private investments through 2030, assuming no new government policies are implemented. This scenario was defined as the business-as-usual (BAU) scenario.[8]

In Part 3, experts were asked to recommend a total annual federal RD&D budget for the technology area in question. The experts were asked to allocate their recommended budget among basic research, applied research, development, and demonstration investments for specific technologies within the general class of technologies being assessed (e.g., oxy-fired carbon capture technology was one specific technology in the fossil elicitation). Experts were also asked to indicate potential coordination with other areas of energy technology research, as well as industries that could provide "spillover" innovations. In most surveys, experts were also asked to provide their insights into the technological hurdles that could be overcome by research in the areas where they recommended the largest investments, and to recommend research areas for cooperation with other countries.

In Part 4, experts were asked to update their BAU 2030 technology cost and performance estimates under three scenarios of RD&D funding. These scenarios were defined in relation to the level of RD&D recommended by the experts. First experts estimated 2030 cost and performance metrics assuming their recommended annual federal RD&D budget was implemented and held constant over the next 20 years. Then, experts updated their 2030 estimates under two additional RD&D scenarios: a 50% proportional reduction in their recommended budget and a 10-fold proportional increase in their recommended budget. Having experts re-estimate future technology cost and performance was a crucial part of this analysis because it allowed us to estimate the sensitivity of technology outcomes to RD&D funding.

In Part 5, which was not formally a part of all the elicitations we conducted, experts were asked to estimate deployment levels that could be achieved under the four RD&D budget scenarios. These elicitations also asked experts to think through deployment policies that would contribute to commercializing novel energy technologies. In those surveys where deployment was not examined in a separate section, experts were asked similar questions about deployment in Parts 2 and 4.

[8] The business-as-usual (BAU) scenario in this work includes all current deployment policies modeled in the Energy Information Administration's (EIA) *Annual Energy Outlook 2010* (EIA, 2010a). Anadon et al. (2011) summarizes the status quo policies in the BAU scenario and other future policies categorized into power standards, building codes, transportation policies, and climate policies. It also describes the scenarios used to estimate the impact of oil and gas prices.

Four of the elicitations (the bioenergy, fossil, storage, and vehicles surveys) were conducted using a written device, which was mailed to participants. The remaining three elicitations (the nuclear, buildings, and solar surveys) were conducted online. Our three online expert elicitations represented an important innovation, as they were the first formal elicitations about future technology cost and performance to be conducted online. Online elicitations improve the ability of experts to modify their answers and to visualize them as they input their estimates. We included several graphics that allowed the experts to see the uncertainty ranges they specified as well as their estimates of cost and performance under different budget scenarios alongside each other. Online elicitations also accelerated the data collection and analysis process, which would be beneficial for future elicitations conducted on a more frequent or broader scale.

2.5.1. Limitations of this Application of Expert Elicitations

The results presented in the elicitations provide insight into U.S. expert opinion between July 2009 and December 2010 on the states of selected technologies in 2030 and their dependence on federal RD&D investments. However, there are several limitations to an expert elicitation.

Experts are human and can have the same biases toward familiar subjects and overconfidence in their knowledge of unfamiliar subjects. This can have two effects. Experts may overestimate the probability that advances in their particular research area will take place because of availability heuristics, that is, if one can think about a solution one can overestimate the probability that the solution will materialize (Tversky & Kahneman, 1973). One could also argue that experts may have a motivational bias to overestimate the required federal RD&D investments and their returns if they stand to personally gain from increased federal RD&D funding. However, we see no significant evidence that experts favored their own areas of expertise when recommending a budget allocation. It is also possible that some experts, aware of projections in the literature (which were also included in the background information of the surveys), anchored their estimates to them.

Our study was also limited in scope to focus on federal RD&D investment recommendations and outcomes. We asked experts to estimate the impact of their recommended federal RD&D program on technology costs and performance in 2030 assuming that only the federal budget changes as they recommend. *Cost and performance, however, are shaped by both public and private RD&D funds.* There is little empirical analysis of the interaction between public and private RD&D funds for energy technologies that we could have utilized to quantify the impact that an enhanced federal energy technology innovation program would have on private sector activities. It is possible that experts considered private sector RD&D activity that would be induced by the RD&D activity funded by their recommended RD&D budgets when completing the elicitation, but we did not explicitly ask experts about how they thought private sector activity would change in response to federal RD&D programs.

Expert elicitation is different from surveys in that individual respondents are not treated as observations from a single population. Instead, expert elicitation treats individual respondents as representatives of a large body of knowledge. Further, the emphasis of developing a "population" of experts to participate in an expert elicitation is on quality and diversity of expertise, not on the quantity of participating experts (Cooke, 1991). The Delphi process or expert consensus methods (such as those used by the reports of the Intergovernmental Panel on Climate Change) are closely related to the methodology of expert elicitations. In the Delphi process, expert judgment is elicited in a group and the emphasis is to reach a consensus judgment (Dalkey, 1969). In expert elicitation, each expert's judgment is elicited individually. The advantage of expert elicitation is that each expert offers his or her judgment without being influenced by other experts, avoiding unwanted group dynamic effects obscuring the true diversity of judgments. But expert elicitation also has disadvantages compared to the Delphi process, namely that developing a coherent message from expert judgment is left up to the analyst who ultimately must decide how to weigh assessments of different experts against each other. Because of this, we complemented the elicitations with qualitative interviews in which we presented the elicitation results to a set of 23 stakeholders who were not involved in the first round of elicitations but had expertise managing RD&D budgets for each technology area. We refer to these stakeholders as "program managers" from now on because they have expertise in thinking about investments in technology portfolios, whether they are venture capital or private equity investors, U.S. Department of Energy program managers, or U.S. Congressional staffers. This second set of experts helped interpret the answers of the elicitations. We conducted two to six consultations with program managers for each of the seven elicitations either in person or over the phone (a total of 23 consultations). In these conversations, which lasted from one to two hours, we showed the program experts the technology area experts' recommended budgets and technology cost and performance parameter estimates. The program experts were asked to give their reactions to the budget allocation recommendations and technology cost and performance forecasts. Their insights were used to select three sets of expert assessments per technology area representing pessimistic, optimistic, and middle-of-the-road technology trajectories to be examined in MARKAL. By distilling the results of the expert elicitations to just three trajectories that spanned the range of the full set of assessments from more than 100 experts, we were able to address the computational difficulty that would have arisen if we had attempted to simulate every expert's assessment individually. In addition, it was not necessary to conduct scenario analysis with every combination of experts because defining these three types of technology profiles would essentially bound the range of benefits of RD&D.

2.5.2. From Expert Elicitation Responses to Model-Ready Scenarios of Technology Cost Trajectories

From the expert elicitations we obtained 10th, 50th, and 90th percentile estimates of technology costs in 2010 and 2030 under different federal energy RD&D budget scenarios. To understand how these technology improvements are likely to affect the uncertainty around CO_2 emissions, carbon prices (in carbon policy scenarios), and oil imports between 2010 and 2050, we translated these estimates into the multiregion U.S. MARKAL model. To do this, we first created full probability distributions of technology costs. We then interpolated between the 2010 and 2030 estimates, and extrapolated from 2030 to 2050. To represent the variance around the metrics we care about, we sampled across these technology distributions utilizing a computationally efficient strategy that allowed us to sample across the distributions of the different technologies, accounting for dependence across technology improvements. This process of creating inputs to MARKAL was applied to four RD&D budget scenarios (BAU, half of recommended budget, recommended budget, and 10-times the recommended budget) and three types of experts (pessimistic, optimistic, and middle-of-the-road), and is described in detail in the next sections. For more detail, readers are referred to Anadon et al. (2011).

2.5.3. Energy System Model and Policy and Market Scenarios

The MARKAL model is used to understand the interaction of the energy technology areas covered by the elicitations in the context of the U.S. economy. MARKAL is a technology-based, bottom-up model with flexibility to add novel technologies that is widely used by federal decision makers. DOE and EPA have each developed their own versions of MARKAL for policy analysis. We utilized a revised version of the U.S. multiregion MARKAL model developed by Brookhaven National Laboratory and used by DOE. This version of MARKAL models the United States as 10 separate regions and includes trade between regions to capture the differences in energy resources and investment and the nature of energy demand in each of the regions. MARKAL is a partial equilibrium model that is specifically designed to represent technological evolutions of the physical energy system, occurring over 30- to 50-year periods. As a linear optimization model, it is solved as a cost minimization problem where future states of the energy system are determined by identifying the cost-effective pattern of resource use and technology deployment over time, given exogenously specified energy demands. The MARKAL objective is thus to minimize the total cost of the system, discounted over the planning horizon. Each year, the total cost includes: annualized investments in technologies; fixed and variable annual operation and maintenance (O&M) costs of technologies; cost of exogenous energy and material imports and domestic resource production (e.g., mining); revenue from

Table 2.1. *Scenarios analyzed in this work*

Technology cost assumption Profiles	• Middle-of-the-road • Optimistic • Pessimistic
RD&D levels	• Business-as-usual • Half recommendation • Full recommendation • Ten times recommendation
Deployment policies	• Current policies only • Economy-wide CO_2 cap-and-trade program • Sectoral energy standards for electricity, commercial buildings, and vehicle fuel economy
Energy prices	• Annual Energy Outlook 2010 (AEO2010) reference case prices • Low gas price (AEO2011 reference case) and reference case oil price (AEO2010) • High oil price and high gas price (AEO2010)
Nuclear power expansion	• Available and deployed where economic • Unavailable due to unforeseen constraints

exogenous energy and material exports; fuel and material delivery costs; welfare loss resulting from reduced end-use demands; and taxes and subsidies associated with energy sources, technologies, and emissions. For this analysis, the endogenous technology-learning and stochastic programming features of MARKAL were not used; this allowed us to specify exogenous technical change based only on the results of the elicitations.

We use a number of scenarios in the MARKAL model to explore the interactions of various policies and market variables with RD&D portfolios (see Table 2.1). Table 2.1 shows the composition of the scenarios that we analyzed in this study.

The default assumptions in the scenarios are based on the *Annual Energy Outlook 2010* (EIA, 2010a). In particular, these assumptions include crude oil and natural gas prices, as well as existing deployment policies, but they exclude technology cost assumptions as these are taken from the expert elicitations.

The scenarios with an economy-wide CO_2 cap-and-trade policy are based on the president's goal of 83% reduction from 2005 GHG emissions by 2050. Banking and borrowing are allowed, as are domestic offsets. Scenarios with new sectoral policies include a federal clean-energy standard, commercial building standards, and vehicle fuel economy standards (Corporate Average Fuel Economy [CAFE]). The federal clean electricity standard requires clean electricity generation to reach 45% by 2015, 80% by 2035, and 95% of sales by 2050 with no special exemptions from the baseline. Generation from natural gas qualifies for a 50% clean-energy credit; coal with CCS,

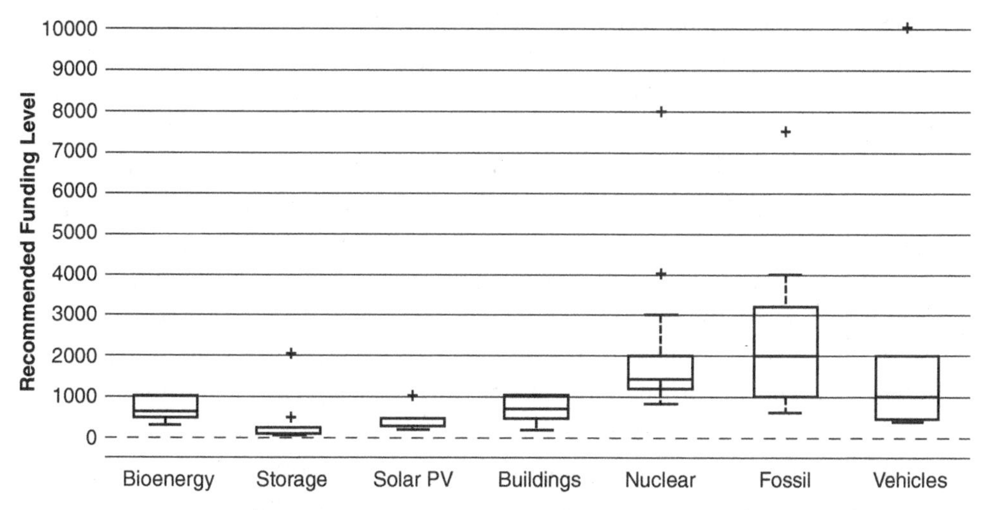

Figure 2.4. Range of budget recommendations from experts by technology area. The horizontal line in the rectangle represents the median recommendation; the bottom and top of the rectangular box correspond, respectively, to the 25th to 75th percentiles; and the discontinuous line represents the highest and lowest recommendations within 1.5 times the interquartile range of the budget recommendations. The recommendations outside this range are shown in cross marks.

90%; and natural gas with CCS, 95%. The increased CAFE standard assumes a linear increase from 35 miles per gallon in 2020 to 75 miles per gallon by 2050. Commercial building standards result in a 30% improvement in building shell efficiency.

Sensitivities to oil and natural gas prices are explored to estimate the sensitivity of RD&D investments to low natural gas prices (a likely possibility given the recent development of shale gas resources in the United States) and to high fossil fuel prices. We parameterize the energy price scenarios using (1) the Annual Energy Outlook (AEO) 2010 reference case oil price and a low gas price case based on the AEO 2011 reference case (EIA, 2010b); and (2) the AEO 2010 high oil and gas price cases.

2.6. What Do Experts Think about the Future of Energy Technologies and the Role of Government RD&D?

2.6.1. Recommended Public RD&D Investments

The large majority of the more than 100 experts who participated in the study recommended that government RD&D investments in each of the technology areas that we studied should be increased above current levels. However, the degree to which experts recommended RD&D funding be increased varied significantly. Figure 2.4 shows the distribution of recommended public RD&D investments by technology area. The average recommended increases over fiscal year (FY) 2009 appropriations ranged from 186% for solar photovoltaic to 963% for utility-scale storage (where the base for comparison is small and the potential is large). In absolute terms, the average recommended increases in spending ranged from $221 million per year for storage to $1.65 billion per year for fossil energy.

On average, the 12 *bioenergy* experts recommended $680 million in annual spending on bioenergy RD&D, representing a threefold increase over 2009 spending on bioenergy RD&D ($220 million). Experts placed a slight emphasis on pilot and commercial demonstration projects, allocating on average 20%, 22%, 29%, and 29% of their recommended budgets to basic research, applied research, experiments and field demonstrations, and commercial demonstrations, respectively. When allocating their recommended budgets among technologies, the greatest fractions of the total budget were allocated to gasification, hydrolysis, and "other" research that most experts specified as feedstock development, harvesting, and transportation.

On average, the 25 *storage* experts recommended $240 million of annual spending from 2010–2030 on grid-level energy storage RD&D. This is a fourfold increase compared to FY 2010 spending of $63 million by DOE. Experts emphasized later stages of RD&D slightly more than early stages such as basic research. Commercial demonstration received an average of 31% of the budget; experiments and field pilots, 26%; and applied research, 24%. Basic research was allocated 19%. According to the experts, the majority of this budget should be spent on flow batteries, batteries, and compressed air energy storage (CAES), which account for 24%, 18%, and 18% of the average recommended budget, respectively. They believed that flow batteries were close to maturity, allocating 16% of the average total budget to experiments and pilots or commercial demonstration. Batteries were deemed to need equal amounts of funding at all four stages, and CAES primarily at the commercial demonstration stage.

The average recommendation for the optimal level of federal RD&D funding for *nuclear* energy over 30 experts was $1.8 billion per year. No expert recommended maintaining or reducing RD&D funding from the 2009 level of approximately $466 million per year (an amount that excludes investments in facilities of around $300 million per year). Experts allocated the greatest amounts of spending to very high temperature reactors (note that some experts indicated that research on reactors operating at around 800°C should also be included in this category), fuel cycle, sodium-cooled fast reactors, fuels, and small and medium factory-built modular reactors. When looking at how experts allocated their spending on a percentage basis across different innovation stages, experts allocated on average 31% of funding to applied research, almost 30% to experiments and pilots, and around 20% for basic research and for commercial demonstration.

The average recommendation for the optimal level of RD&D funding among the 13 *fossil energy and CCS* experts was $2.3 billion per year. No expert recommended maintaining or reducing RD&D funding from 2009 levels (approximately $400 million per year). Experts placed strong emphasis on commercial demonstration, on average recommending that more than 40% of the total RD&D budget for fossil energy be allocated for this purpose (the maximum and minimum fractions experts dedicated for demonstrations are 80% and 15%, respectively). Within commercial demonstration, experts emphasized integrated gasification combined cycle (IGCC), oxy-fired combustion, and retrofitting existing plants for CCS.

The nine experts in *building technology* recommended an average of $680 million per year in RD&D, a more than fourfold increase over FY 2010 spending of $144 million. On average the greatest amounts of spending were allocated to applied research for windows (4%) and controls and monitoring (5%), as well as building metering experiments and pilots (5%). Experts generally spread their recommendations relatively evenly across the different categories, and further defined their recommendations for controls and monitoring and building metering to include full system monitoring as well as metering energy efficiency down to a component level. On average, 31% of total funding was directed to commercial demonstrations and 27% for experiments and pilots.

The nine *vehicles* experts we elicited recommended an average of $2.05 billion per year in vehicles RD&D spending. The recommended spending levels were highly skewed; the median recommended RD&D spending level was $1 billion, indicating that, although most experts recommended large increases in RD&D spending from the current level ($485 million in FY 2010), there is large disagreement over the extent to which the vehicles' RD&D program should expand. On average, experts allocated just over 35% of their recommended RD&D budgets to both basic research and applied research, and 20% to experiments and prototypes, with the remaining allocation to commercial demonstration. The technology areas with greatest RD&D emphasis were lithium-ion batteries, novel concepts for energy storage, and materials, each receiving 12–15% of recommended RD&D funds. Ultra capacitors, battery manufacturing processes, hydrogen storage, and electronic controls were also emphasized, each receiving more than 7% of recommended RD&D funds.

The eleven *solar photovoltaic* (*PV*) experts recommended, on average, $410 million per year for RD&D. This is a more than twofold increase compared to FY 2010 spending of $175 million by DOE (including PV, system integration, and market transformation). Experts recommended on average 37% for applied research, and 27% and 25% for basic research and experiments and pilots, respectively. On average the experts allocated the highest levels of funding to concentrator PV (primarily applied research and experiments and pilots), thin film PV (primarily applied research), and novel high efficiency PV (primarily basic research). Experts recommended funding crystalline silicon, thin film, concentrator, excitonic, and novel high-efficiency technologies (among others), but did focus on thin film, concentrator, and novel high-efficiency technologies.

2.6.2. Expected Impact on Future Costs

Figure 2.5 displays the range of cost reductions across all of the technologies we studied in the seven elicitations. Cost reductions are displayed as the ratio of the median estimated 2030 cost under BAU RD&D to the median estimated 2030 cost under the recommended RD&D portfolio. Notable results from the expert elicitation are discussed below, each in terms of this metric:

- The median of the experts' estimated cost reduction over the cost under a BAU RD&D funding scenario in 2030 ranged from no cost reduction in the case of sodium sulfur batteries (for which only a small number of experts provided estimates) to around 35% in the case of residential-scale solar PV and flywheel energy storage technologies.
- The cost reduction catalyzed by federal RD&D funding beyond the BAU scenario is the lowest for vehicle technologies, ranging from a 2% cost reduction for advanced internal combustion engine (ICE) vehicles to a 10% cost reduction for fuel cell vehicles. (This may reflect a judgment that investments by private firms, rather than federal spending, will be the major driver of improvement in vehicle technologies.)
- The figure shows that experts estimate the cost to reduce heating and cooling energy needs in new commercial buildings will be reduced 5% to 52% over BAU costs.
- The benefits of the recommended RD&D portfolio to the various energy storage devices are mixed: experts estimated that hydro storage costs could not be reduced below their BAU estimates; batteries and generic technologies may achieve up to 14% and 25% reductions in cost over the BAU, respectively; and CAES and flow batteries costs will decrease by 7% to 38% and 2% to 58% respectively.
- There was considerable uncertainty among nuclear technology experts about the benefits to be gained by spending on the recommended RD&D level. The most pessimistic experts thought that small modular reactor (SMR), Gen III/III+, and Gen IV costs would not decrease with additional RD&D. But the most optimistic expert estimated that SMR, Gen III/III+, and Gen IV costs could decrease by 38%, 24%, and 24%, respectively.
- Bioenergy experts were in greater agreement on the plausible gains to gasoline substitutes than to any other biomass derived fuel under their RD&D recommendations. They estimated that gasoline substitute costs would be 51% to 69% lower than BAU costs. Experts also estimated that major cost reductions, 27%, 86%, and 56% respectively, could be achieved for electricity, jet fuel substitute, and diesel fuels derived from biofuels.

2.7. From Technology Costs to Societal Outcomes: Accounting for Technology Uncertainty

In this section, we present and discuss some example results from using the expert elicitation data in the MARKAL model to investigate the relationship between public RD&D in energy and societal outcomes. The outcome metrics we present in this analysis include energy-related CO_2 emissions, oil imports, CO_2 prices, and CES

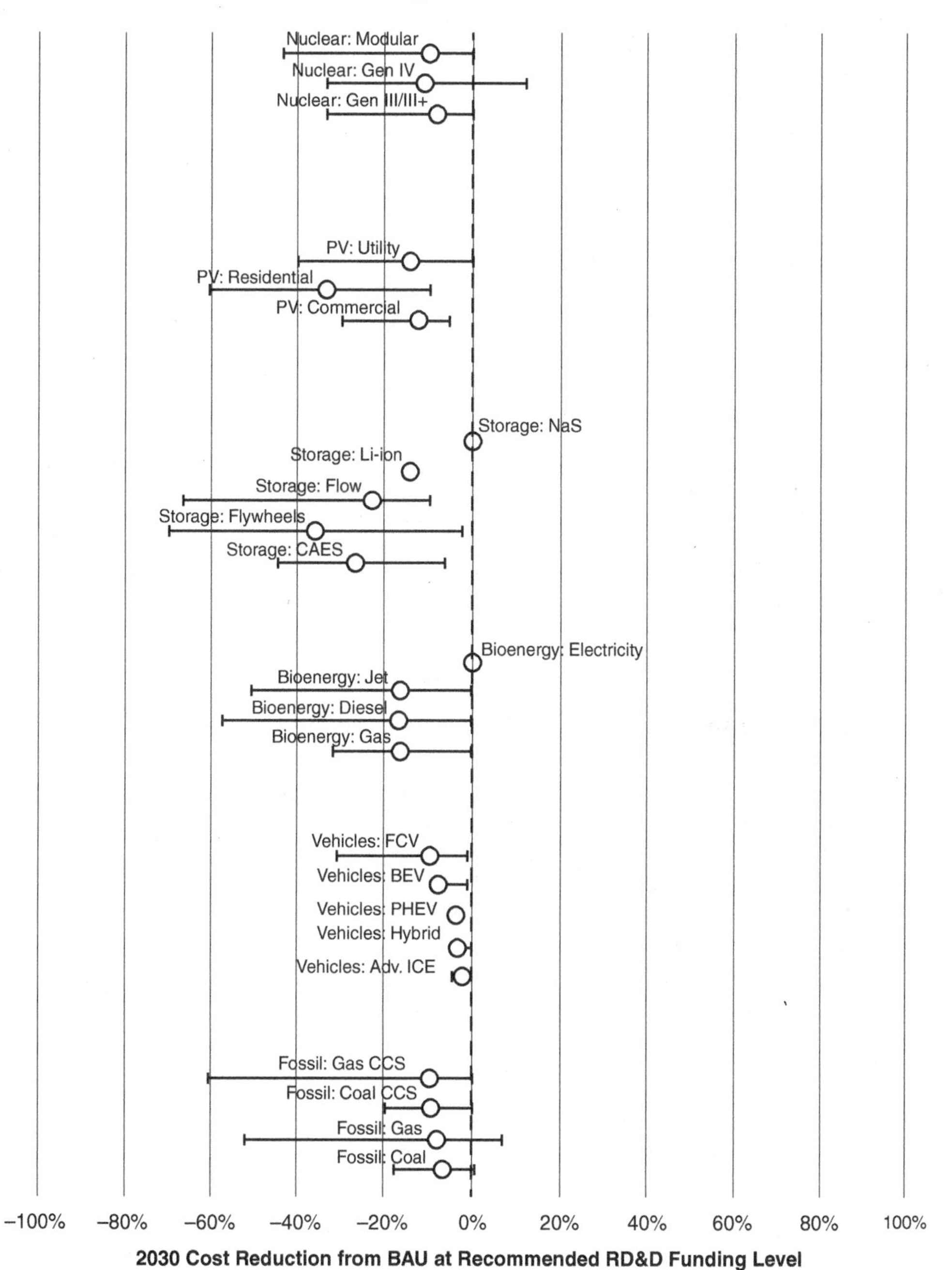

Figure 2.5. Range of median cost reduction per technology predicted by each expert under his or her recommended RD&D spending per industry. Costs cover storage technologies; overnight capital costs for grid level energy; overnight capital costs for SMR, Generation IV, Generation III/III+ nuclear power; production costs of biomass-based electricity ($ per kWh) and substitutes for jet fuel, diesel, and gasoline ($ per gallon); unit costs for advanced internal combustion engine (ICE), hybrid, plug-in hybrid, electric, and fuel cell vehicles ($ per unit); overnight capital cost of coal and natural gas power plants with and without CCS ($ per kW); and module cost for residential, commercial, and utility scale PV panels ($ per kW).

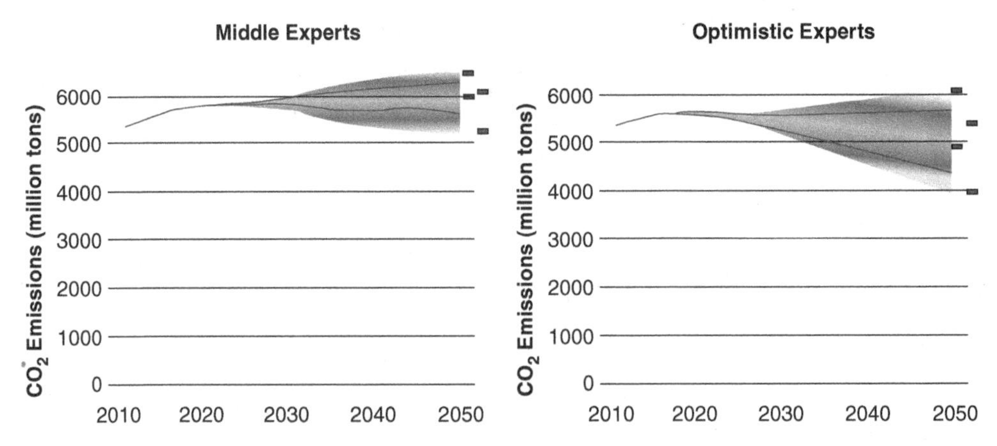

Figure 2.6. U.S. energy-related CO_2 emissions under BAU federal RD&D investment with no additional demand-side policies (top cloud marked by tick marks closest to the figure on the right) and 10-times the experts' average recommended federal energy RD&D investments (between $49 and $82 billion per year) (bottom cloud marked by the tick marks furthest from the figure on the right) with no additional demand-side policies using middle-of the-road (left) and optimistic (right) experts' technology cost projections.

credit prices. We show how the results change in the different policy scenarios and with different inputs from the elicitations.

As discussed earlier, to investigate the impact of different levels of RD&D funding for each of the seven technology areas, we developed three types of technology assumptions from the experts' responses: "middle-of-the-road," "optimistic," and "pessimistic." To simplify the presentation of sample results, we summarize results from the middle and optimistic technology assumptions and we do not refer to the modeling results under the pessimistic technology assumptions.

A robust result of this approach is that even very large energy RD&D increases are not expected to result in large emissions reductions to achieve climate mitigation goals by themselves. These results illustrate the need for additional policies working in conjunction with RD&D to reduce CO_2 emissions from the energy system. For example, projections from the middle-of-the-road experts suggest that even large increases in energy RD&D funding, in the absence of a limit on carbon emissions, still lead to a roughly 95% probability that emissions in 2050 would be greater than those in 2010, far from the Obama Administration's goal of a more than fivefold reduction. The uncertainty shown in Figure 2.6 includes only the experts' estimates of the uncertainty around the cost and performance of the seven technology areas covered in our surveys, not the myriad other factors that will affect U.S. emissions between now and 2050. Under a no-new-policy scenario, CO_2 emissions continue to rise from today to 2050 under BAU funding levels. Emissions level off and decline under full and 10-times scenarios under both middle and optimistic technology assumptions;

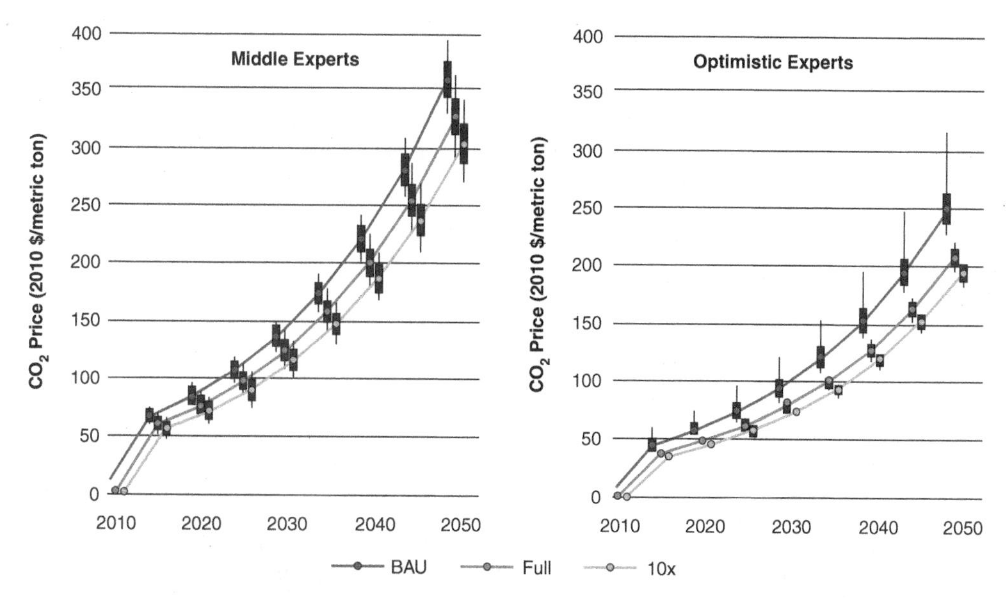

Figure 2.7. Trajectory of CO_2 prices ($ per metric ton of CO_2) under a CO_2 cap limiting emissions to 83% of 2005 levels in 2050, with no international offsets (note that this is a very stringent policy). Three RD&D funding scenarios are shown: BAU (top curve), full recommended (middle curve), and 10-times the recommended level (bottom curve). The graph on the left includes the results using the technology cost assumptions from the middle-of-the-road experts and the graph on the right includes the costs from the optimistic experts. The dot corresponds to the 50th percentile, the black rectangle encompasses the 25th and 75th percentiles, and the line encompasses the 5th and 95th percentiles.

however, they are still at levels that would make it impossible to achieve climate goals set by the United States in international fora for mitigating dangerous anthropogenic climate change.

Figure 2.7 illustrates that, while a CO_2 cap alone could reduce emissions, RD&D would make it significantly less expensive to do so, reducing the CO_2 allowance price needed to meet the cap. With the middle technology assumptions, median CO_2 prices in 2050 fall from more than $360 per ton of CO_2 under the BAU scenario, to around $330 per ton under the recommended RD&D portfolio, and to just over $300 per ton under the 10-times RD&D funding scenario. Under the optimistic technology assumptions, the CO_2 price declines with higher levels of RD&D funding and the uncertainty in the price also decreases.

In the scenario without new climate policies, oil imports are generally expected to decline until 2035, due mainly to existing vehicle fuel economy standards and other existing policies and trends (Figure 2.8). Under the assumptions of the middle experts and absent other policies, increasing levels of funding from BAU to the full recommended portfolio to the 10-times the recommended portfolio reduces the median

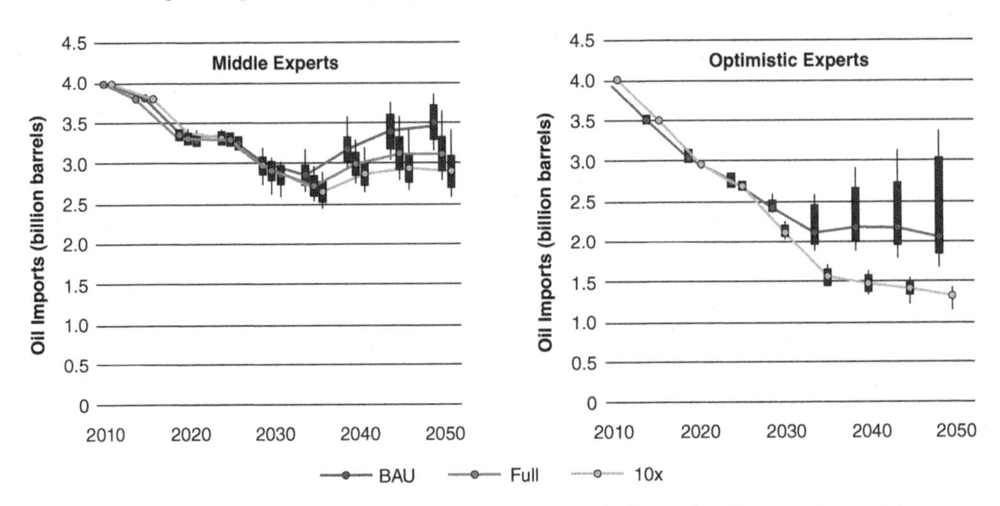

Figure 2.8. Comparison of oil imports per year (excluding other imported petroleum products) under a no-policy case by RD&D level and type of assumptions. In the graph on the right the full recommended funding case would fall between the BAU and 10-times RD&D funding levels.

projected level of oil imports, especially post-2035. Under optimistic assumptions and 10-times funding levels, oil imports continue to decline to 2050. Increased levels of funding without new energy policies are insufficient to reduce U.S. oil imports by one-third by 2021 (a goal stated by President Obama in March 2011). It is notable that the uncertainty around oil imports after 2030 in the optimistic assumptions for the BAU RD&D scenario becomes very significant, showing that a high degree of technology uncertainty can greatly affect oil imports.

2.7.1. Comparison of Economic Performance of Different Scenarios

Under all the scenarios modeled, consumer and producer surplus increases and total energy system cost declines with increases in RD&D funding. We find that an investment of $5 billion dollars per year in energy RD&D funding instead of current investment levels ($2.1 billion) would result in about $75 billion per year in increased consumer surplus[9] by 2050 in a scenario with no additional policies. This increase is even larger under the optimistic technology experts' assessments (Figure 2.9). Under the CES case and middle assumptions, increasing applied energy RD&D from $2.1 billion to about $5 billion would result in an increase in consumer and producer surplus of $350 billion in 2050. Under the CO_2 case and middle assumptions, the

[9] Note that, as a measure of impact of RD&D, consumer and producer surplus is a metric that encompasses more factors than other studies measuring the returns to RD&D, such as the National Academies report (NRC, 2001), but does not consider the opportunity cost of investing the energy RD&D funds elsewhere in the economy, nor does it consider other benefits that may result from the reduction of CO_2 emissions.

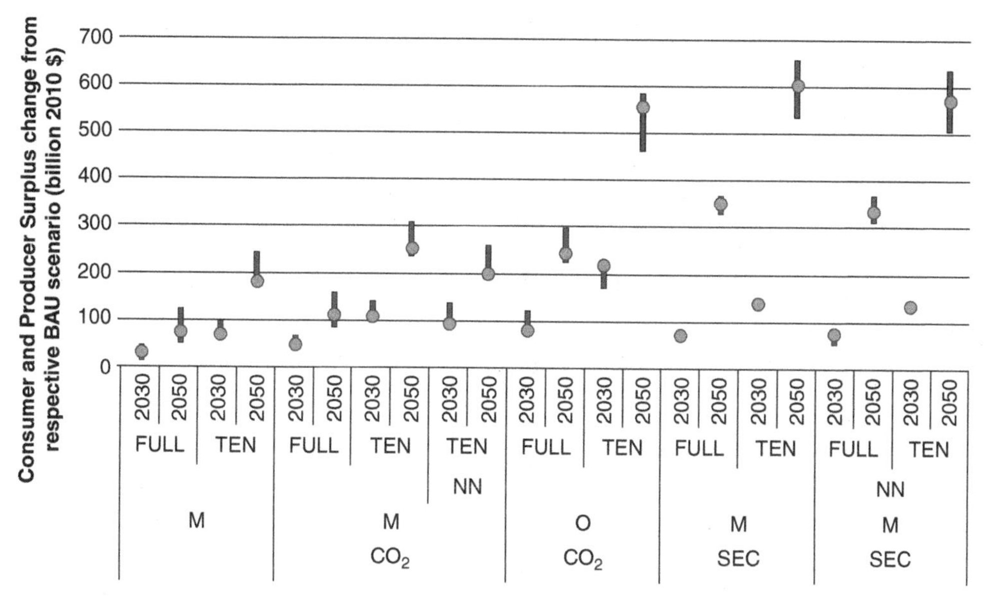

Figure 2.9. Consumer and producer surplus projections over the respective BAU RD&D scenario in 2030 and in 2050. The dot corresponds to the median result, and the edges to the 5th and 95th percentile results. FULL = recommended RD&D scenario; TEN = 10-times recommended RD&D scenario; CO_2 represents results for the CO_2 cap scenario, and CES the results for the sectoral policy scenario.

increase in consumer and producer surplus in 2050, resulting from increasing energy RD&D from $2.1 billion to about $5 billion, is about $110 billion.

2.7.2. Evidence of Substantial Returns to RD&D Investments – Eventually Decreasing

Most of the experts we surveyed expected decreasing marginal returns[10] to energy RD&D investment in terms of technology cost reductions and performance improvements; this result from the elicitations was reflected in our modeling results. For example, Figure 2.10 shows the change in consumer surplus per change in the level of RD&D investment when increasing RD&D funding from the BAU level to the full recommended level and subsequently from the full recommended level to the 10-times recommended level under the CO_2 cap policy scenario. The median result shows that

[10] The persistent finding that experts projected decreasing marginal returns to energy RD&D investments reflects the finite number of available RD&D projects that could be feasibly conducted in the short run. Over the long run, the pool of available RD&D projects will evolve due to the cumulative nature of innovation: today's successful RD&D projects push out the "technology frontier," opening up new avenues for future RD&D investments. Therefore, it does not follow that the projected decreasing marginal returns to RD&D in our results imply that future RD&D investments will have lower returns than today's investments. The continuously advancing technology frontier highlights the need for any empirically driven approach to informing RD&D investment to be continuously updated every few years.

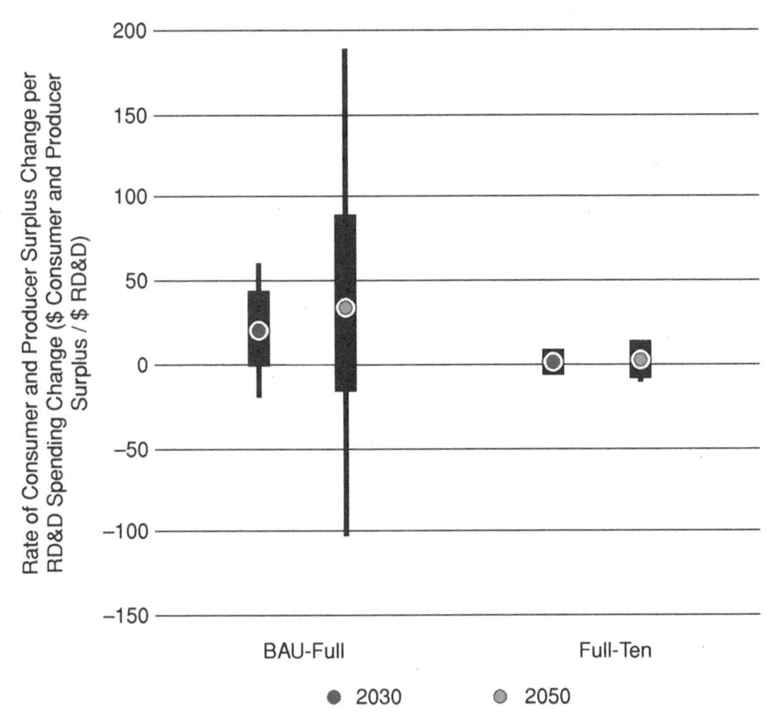

Figure 2.10. Average increase in consumer and producer surplus per dollar of RD&D invested when comparing the 2030 (to the left in each of the two scenarios BAU-Full and Full-Ten) and 2050 (to the right in each of the two scenarios BAU-Full and Full-Ten) results of increasing RD&D investments from the BAU to the full recommended scenario (two bars on the left) and from the full recommended to the 10-times recommended scenario (two bars on the right), all under the 83% CO_2 reduction cap. The dot corresponds to the 50th percentile, the black rectangle encompasses the 25th and 75th percentiles, and the line encompasses the 5th and 95th percentiles.

increasing RD&D funding from the BAU to the full recommended scenario under a CO_2 cap policy results in an increase in consumer surplus in 2050 of about $33 per dollar invested in energy RD&D whereas the corresponding change increasing from the full recommended level to the 10-times recommended level results in only a $3 return to consumer surplus per dollar invested in energy RD&D.

This analysis suggests that by using the input we obtained from the seven expert elicitations, from the multiregion MARKAL model, and from our sampling strategy, that an increase of at least $10 billion in the RD&D investments in the seven energy technology areas we considered would have positive social benefits. Although the marginal returns to RD&D investment lead to smaller returns at higher levels of investment, even at the highest range we consider, the 2050 benefits of RD&D investment outweigh the costs. Thus, our methodology, if expanded to a broader range of hypothetical RD&D budgets, could be used to estimate the optimal level of RD&D investment to maximize social benefits.

In 2011, based on a consideration of estimated economic benefits, direct expert recommendations, and political constraints, we recommended an increase of U.S. government energy RD&D funding for the six technology areas that we modeled plus buildings (which we covered using just the expert elicitation) from $2.1 billion to $5.2 billion. We made an overall recommendation of $10 billion that included $5.2 billion for the seven energy technology areas, funding for energy RD&D in the other DOE program areas (e.g., industry, wind, geothermal), ARPA-E, and Basic Energy Sciences (Anadon et al., 2011). Further modeling for this book and other work (Chan & Anadon, 2014) indicates that based on a consideration of only economic benefits, even higher levels of funding, amounting to $10 to $15 billion for the six areas we modeled plus buildings would have positive overall benefits for society. This would imply a total of $15 to $20 billion for all energy RD&D programs (not just the seven we covered in our overall analysis), ARPA-E, increase in funding than we called for in 2011, though of course this increase would be phased in over a number of years.

2.8. Optimizing the Allocation of Energy RD&D Investments under Different Scenarios

In the previous section we discussed how one can use the results from expert elicitations to estimate the optimal level of RD&D investments, defined as the RD&D investment level after which selected measures of benefits exhibit marginal returns that do not justify their costs. In this section we develop and apply a methodology for relating technology costs and model output to RD&D portfolios, allowing the estimation of optimal RD&D portfolio allocations. This analysis accounts for uncertainty in technology costs given RD&D, substitution and complementarities among technologies, nonlinearity in the relationship between RD&D and technology cost, and multiple (although not simultaneous) policy objectives.

The initial stage of the approach requires us to develop a method to "invert" the expert elicitation and cost sampling process, estimating RD&D levels given technology costs rather than technology costs given RD&D levels. The second stage of the approach fits a multidimensional surface to RD&D levels and an expected outcome parameter from an importance sampling mapping based on output from the MARKAL model (such as the CO_2 emissions allowance price, CES price, CO_2 emissions, total system cost, economic welfare, etc.). Finally, the last stage optimizes over this surface to minimize (or maximize) the outcome variable under a total budget constraint. (For more details, readers are referred to Anadon et al., 2011 and Chan & Anadon, 2014.)

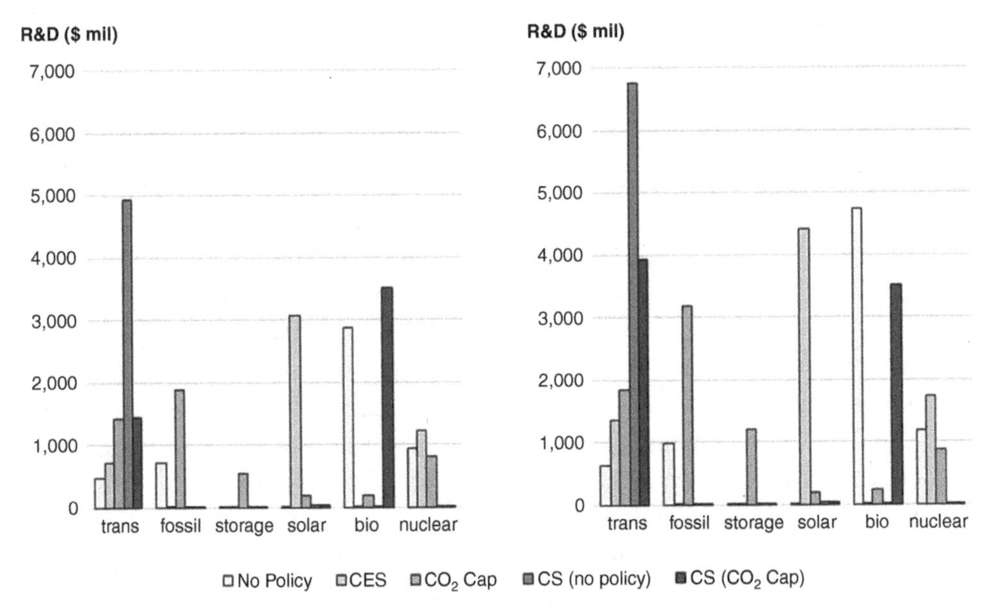

Figure 2.11. Optimized budget allocation under different scenarios: no-policy optimized for total CO_2 emissions, standards policy (CES) optimized for CES credit price; CO_2 cap-and-trade (CO_2 cap) optimized for CO_2 price; no-policy, optimized for consumer surplus (CS [no policy]); and CO_2 cap-and-trade optimized for consumer and producer surplus (CS [CO_2 cap]). The left figure optimizes these scenarios under a $5 billion budget and the right figure optimizes the same scenarios under a $7.5 billion budget.

We now show example results calculated using this approach applied to the expert elicitation results described in the preceding text. Using an estimated seven-dimensional surface relating the technology-area RD&D levels[11] with the expected outcome metric, we calculate optimized portfolios using a constrained nonlinear optimization algorithm. The function defining the seven-dimensional surface is minimized (or maximized) along the dimension of the outcome metric, allowing the six RD&D levels to vary, but constraining the six RD&D levels to have a sum less than a specified total budget amount. This approach represents the RD&D investment decision problem as a one-stage decision to set the level of RD&D investment; it is just one of various options to calculate the optimal allocation of RD&D investments from expert elicitations and the outcomes of energy economic models.

Using this methodology we estimate the optimum RD&D portfolio for three policy cases: no-new-policy, CO_2 cap-and-trade, and a CES. The RD&D investment portfolio in the no-new-policy case was optimized to minimize CO_2 emissions and to maximize consumer and producer surplus (both in Figure 2.11). In the CO_2 cap-and-trade

[11] We drop building technologies from this analysis because this elicitation did not elicit uncertainty ranges in cost parameters.

case, the RD&D portfolio was optimized to minimize CO_2 prices and to maximize consumer and producer surplus. As an example we show the optimal RD&D investments for these five cases for two total energy RD&D budgets (or budget constraints) of $5 billion per year and $7.5 billion per year. These results show clearly that the optimal level of RD&D of many technology areas varies depending on the policy and optimization metric and, to a lesser extent, on the budget constraint.

Because the optimal allocation of RD&D investments to different technology areas is contingent on the policy scenario, and because it is often far from clear which policies will be implemented in the future, a diversified portfolio provides option value.

As with any other modeling approach, this methodology has limitations. First, future technological breakthroughs are inherently unpredictable, and the experts we surveyed inevitably have limited foresight. The uncertainties we are able to incorporate, for example, are only the uncertainties judged by the experts (and the variations among them); the real uncertainties in outcomes may be much bigger than the experts perceive them to be.

Second, our economic model inevitably incorporates a wide range of assumptions, some of which may prove to be wrong. Some factors are clearly absent from the model, such as the increase in private sector innovation investment that would arise under different policy scenarios. Another important element, not included in our models, is the likelihood that if the United States (and other countries) succeeded in using new technologies to reduce their oil consumption, global oil prices would be reduced, with significant energy security benefits that are difficult to quantify.

A third factor has to do with the sampling approach used in this section to optimize the portfolios, which leaves out the option value of funding a technology with uncertain benefits. While we capture the uncertainty in the relationship between RD&D levels and outcome variables in the first two stages of our approach, once the surface over the RD&D levels and the outcome variable is calculated, all uncertainty drops from the model in the optimization step. A more sophisticated approach could propagate uncertainty into the optimization stage using a stochastic optimization algorithm at the cost of additional computational power. The consequence of the approach we take is that the option value of RD&D investments is not accounted for. Therefore, for technologies with RD&D that affect future technology costs with high variance, RD&D in this technology has high option value, but that will not be reflected in the expected value considered here. Hence, our method relatively understates the value of RD&D in technology areas with greater uncertainty in future costs.

Fourth, there are a number of other benefit vectors that MARKAL is not able to incorporate in its outputs. The potential for clean-energy investments to create large numbers of well-paying jobs is clearly an important factor, but given the complexities in estimating how many jobs would be created decades in the future by one investment versus another, we did not attempt to incorporate this issue into our modeling. Similarly, our modeling did not address what fraction of the global market for energy technologies – estimated at tens of trillions of dollars over the next four decades – the United States would be able to supply, in competition with other countries, but it seems clear that increased energy RD&D will be an important factor in determining what role the United States will play in global energy technology markets. We also did not attempt to model the benefits for U.S. national security of reduced dependence on oil imports or reduced military dependence on vulnerable fuel supply lines. Perhaps most importantly, we modeled only technology deployment in the United States, though many of the technologies that might be developed in the United States would have their biggest impacts on reducing carbon emissions or global oil use through their deployment in other countries. In all these respects, the modeling in this chapter almost certainly understates, rather than overstates, the case for additional investment in RD&D.

2.9. Broader Recommendations for the Energy RD&D Decision-Making Process

This chapter first describes the history of funding for energy RD&D by the largest public funder in the United States, the U.S. Department of Energy. We have argued that the process for making decisions about energy RD&D investments at DOE should be more transparent, consistent, and should include considerations of uncertainty and market interactions. Although over the past decade different Offices in DOE have put in place a few efforts that have moved in this direction, the progress seems to have stopped over the past few years. We posit that the lack of the type of analysis we propose is due to concerns among the different DOE programs that internal exercises aimed at projecting future benefits considering various technology programs simultaneously leaves them with an insufficient ability to control the final outcome and could ultimately hurt their budgets; and, perhaps to a lesser extent, concerns among external agencies about the validity of modeling assumptions. To address the shortcomings of current decision-making processes, we have proposed utilizing a transparent, consistent, and more holistic analysis that relies on expert elicitations and an energy-economic model, such as MARKAL, to consider the probabilistic social benefits of public RD&D-induced technical change. We also argue that it is important to include analysis considering different future policies, as well as energy prices. The approach we present in this chapter has already been the subject of some interest to policymakers at DOE, as well as in other government agencies throughout the world tasked with making investments on energy RD&D and would provide a complement to current approaches.

Our analysis quantitatively supports a total energy RD&D budget in the range of \$10 to \$15 billion, excluding funding for the Office of Science and ARPA-E – this would represent a significant increase over historical levels (see Figure 2.1). As new technological possibilities are explored, it may be desirable to increase funding even further. It could also be the case that if sufficiently strong carbon policy is put in place – which, we have argued, is essential to reducing emissions and oil imports to levels consistent with the stated goals of the president – after a few years, private actors could undertake sufficient innovation so as to reduce somewhat the need for government investment. However, the impact of a CO_2 cap or sectoral policies (such as those studied in this work) on technology outcomes or private sector innovation levels are unknown, which means that even if such a policy was implemented, it would not be desirable to reduce funding beyond the levels recommended until more evidence is gathered. Furthermore, even greater RD&D investments may be justified, as our work did not cover all technology areas, and technological opportunities may become available in the future that the experts we surveyed were not able to consider.

Our recommended increased investment should be targeted on a broad portfolio of different technologies and different stages of technology, from basic research to large-scale technology demonstrations, and implemented by a portfolio of different institutions appropriate for the different types of innovation (e.g., the Energy Innovation Hubs, the EFRCs, etc.). It is also essential that policymakers understand that innovation is an uncertain enterprise, and that as a result, some failures are to be expected. In fact, if all government programs in energy RD&D succeed, it would almost certainly be the case that the government had not taken sufficient risks – that is, that it was probably supporting many activities that the private sector would be doing anyway.

While the level and allocation of energy RD&D funding is important, it is also important to pay attention to other aspects of the budget-making process. We therefore make the following additional recommendations, based on our work and feedback from experts and stakeholders:

- The coordination between energy RD&D programs and demand-side programs is important. Currently, the U.S. government is making large investments in deployment policies ranging from tax subsidies to loan guarantees to technology standards, often with little coordination with energy research and development programs. To maximize the contribution of federal investments and standards to accelerating energy innovation, these efforts should be closely coordinated, with targeted and appropriate support for each phase of innovation from basic research to initial deployment.
- Federal support should be phased out as technologies become commercially viable, rather than continuing indefinitely. To enable the phase-down of government support for programs or specific projects, the U.S. government should have clear goals and mechanisms to collect and evaluate information about the progress and contribution of projects and technologies from the start. Support programs can also be designed to decrease gradually over time, and to include covenants that allow for program changes based on new information.

- Congress and the administration should work together to design approaches that will provide increased budget stability for energy RD&D programs from year to year. The volatility in spending from one year to the next (shown in Figure 2.2) reduces the efficiency of RD&D programs. The bipartisan cooperation usually maintained in the Energy and Water appropriations subcommittees should be sustained and built on, possibly including the creation of bipartisan task forces focused on particular areas of energy technology.
- Budget decision makers should seek increased input from both technical and market-oriented technology experts from different technology areas. The Quadrennial Technology Review, which was implemented after a recommendation from President's Council of Advisors on Science and Technology (PCAST), has already taken some steps in this direction. The stakeholder meetings with experts from various sectors that we conducted throughout the process are important and useful, especially in improving trust in the process and in gaining valuable insight.
- If expert elicitations are to be used, it is essential to streamline the process as much as possible. This work took approximately four months per elicitation, with the first elicitations requiring seven months. Delays were due primarily to incomplete responses from experts, requiring follow-up phone calls by researchers. Elicitations done online through the Web significantly shortened the process. Interactive, real-time displays of an expert's responses helped the experts correct mistakes immediately, avoiding follow-up phone calls from researchers. Experts also had access to researchers while completing online elicitations if necessary. These features provided a good compromise between interactive tools and access to clarification.
- Going forward, it is important to identify other research areas and other industries that may have technologies that could result in significant reductions in GHG emissions and increased energy security.

Overall, the recommendations and approaches outlined in this chapter can go a long way toward accelerating U.S. energy technology innovation, helping to build a system that meets the CASCADES criteria for an effective energy innovation system outlined in Chapter 1. The analysis in this chapter makes clear that U.S. investments must be *comprehensive*, covering not only RD&D but policies to create incentives for deployment and for private-sector innovation as well. They must be *adaptive*, shifting as some technologies make rapid progress and others do not, or as market conditions change. They must be *sustainable*, as our modeling makes clear that years of sustained investments are needed to achieve the improvements in cost and performance the experts we spoke to project. Our expert elicitation and modeling help identify which combinations of technology investments and demand-side policies would be most *cost-effective*, leading to the most substantial progress toward major goals per dollar spent. Though this chapter did not explicitly address the need for these efforts to be *agile* in responding to the sudden appearance of new opportunities (or realizations that a particular approach is failing), that is addressed in later chapters. Our expert elicitations and modeling make it clear that investments in public RD&D should be *diversified*, focusing on a broad range of technologies and not just attempting to pick a few winners. The approach to estimating the likely benefits of particular RD&D investments outlined in this chapter can help policymakers make these investments more *equitable* between one technology and set of stakeholders and another, providing a common basis to support decisions. Finally, it is clear that these investments must be *strategic*, targeted on achieving key goals such as reduced cost of the energy system, reduced carbon emissions, and increased energy security. Together, these approaches can help build a transformed U.S. approach to energy innovation.

References

Al-Juaied, M., & Whitmore, A. (2009). "Realistic costs of carbon capture." Discussion Paper no. 2009–08. Cambridge, Mass.: Energy Technology Innovation Policy Group. Belfer Center for Science and International Affairs, John F. Kennedy School of Government, Harvard University.

American Energy Innovation Council. (2010). *A Business Plan for America's Energy Future.* Full report. Retrieved from http://www.americanenergyinnovation.org/full-report (Accessed September 20, 2011).

Anadon, L.D., Bosetti, V., Bunn, M., Catenacci, M., & Lee, A. (2012). "Expert judgments about RD&D and the future of nuclear energy." *Environmental Science & Technology*, 41(21):11497–504.

Anadon, L.D., Bunn, M., Chan, G., Chan, M., Gallagher, K.S., Jones, C., Kempener, R., Lee, A., & Narayanamurti, V. (2010). "DOE FY 2011 budget request for energy research, development, demonstration, and deployment: Analysis and recommendations. " Cambridge, Mass.: Energy Technology Innovation Policy Research Group, Belfer Center for Science and International Affairs, John F. Kennedy School of Government, Harvard University.

Anadon, L.D, Bunn, M., Chan, G., Chan, M., Jones, C., Kempener, R., Lee, A., Logar, N., & Narayanamurti, V. (2011, November). *Transforming U.S. Energy Innovation.* Cambridge, Mass.: Report for Energy Technology Innovation Policy Research Group, Belfer Center for Science and International Affairs, John F.Kennedy School of Government, Harvard University.

Anadon, L.D., & Holdren, J.P. (2009). "Policy for energy-technology innovation." In *Acting in Time on Energy Policy*, edited by K.S. Gallagher, pp. 89–127. Washington, D.C.: The Brookings Institution Press.

APS. (2008). "Energy future: Think efficiency. How America can look within to achieve energy security and reduce global warming." American Physical Society. Retrieved from http://www.aps.org/energyefficiencyreport/ (Accessed September 20, 2011).

Baker, E., Chon, H., & Keisler, J.M. (2008). "Advanced nuclear power: Combining economic analysis with expert elicitations to inform climate policy." Social Science Research Network. Retrieved from http://papers.ssrn.com/sol3/papers.cfm?abstract_id=1407048 (Accessed January 18, 2010).

Baker, E., Chon, H., & Keisler, J. (2009a). "Advanced solar R&D: Combining economic analysis with expert elicitations to inform climate policy." *Energy Economics*, 31: 537–49.

Baker, E., Chon, H., & Keisler, J. (2009b). "Carbon capture and storage: Combining economic analysis with expert elicitations to inform climate policy." *Climatic Change*, 96(3):379–408.

Baker, E., Chon, H., & Keisler, J. (2010). "Battery technology for electric and hybrid vehicles: Expert views about prospects for advancement." *Technological Forecasting and Social Change*, 77(7):1139–46.

Baker, E., & Keisler, J. (2009). "Cellulosic biofuels: Expert views on prospects for advancement." Retrieved from http://www.ecs.umass.edu/mie/faculty/baker/CellulosicBiofuels.pdf (Accessed January 18, 2010).

Bird, L., Chapman, C., Logan, J., Sumner, J., & Short, W. (2010). *Evaluating Renewable Portfolio Standards and Carbon Cap Scenarios in the U.S. Electric Sector*. Technical Report NREL/TP-6A2-48258. National Renewable Energy Laboratory. May. Washington, D.C.: U.S. Department of Energy. Retrieved from http://www.nrel.gov/docs/fy10osti/48258.pdf (Accessed February 12, 2014).

Blanford, G.J. (2009). "R&D investment strategy for climate change." *Energy Economics*, 31(1): S27–S36.

Bolinger, M., Wiser, R., Cory, K., & James, T. (2009). *PTC, ITC, or Cash Grant? An Analysis of the Choice Facing Renewable Power Projects in the United States*. Technical Report NREL/TP-6A2-45359. Lawrence Berkeley National Laboratory and National Renewable Energy Laboratory. Washington, D.C.: U.S. Department of Energy. Retrieved from http://www.nrel.gov/docs/fy09osti/45359.pdf (Accessed February 12, 2014).

Bosetti, V., Carraro, C., Duval, R., Sgobbi, A., & Tavoni, M. (2009). *The Role of R&D and Technology Diffusion in Climate Change Mitigation: New Perspectives Using the Witch Model*. Nota di Lavoro 14.2009. Venice, Italy: Fondazione Eni Enrico Mattei. Retrieved from http://www.feem.it/userfiles/attach/Publication/NDL2009/NDL2009-014.pdf (Accessed on February 12, 2014).

Brown, M.A. (2001). "Market failures and barriers as a basis for clean energy policies." *Energy Policy*, 29:1197–1207.

Brown, M.A., Southworth, F., & Stovall, T.K. (2005). *Towards a Climate-Friendly Built Environment*. Report. Washington, D.C.: Pew Center on Global Climate Change. Retrieved from http://www.c2es.org/docUploads/Buildings_FINAL.pdf (Accessed on February 12, 2014).

Chan, G., Anadon, L.D., Chan, M., & Lee, A. (2010). "Expert elicitation of cost, performance, and RD&D budgets for coal power with CCS." *Energy Procedia*, 4: 2685–92.

Chan, G., & Anadon, L.D. (2014). "Improving decision making for public R&D investments with an example in energy." Submitted.

Clarke, J., Weyant, J., & Birky, A. (2006). "On the sources of technological change: Assessing the evidence." *Energy Economics*, 28:579–95.

Cooke, R.M. (1991). *Experts in Uncertainty: Opinion and Subjective Probability in Science*. New York: Oxford University Press.

Craig, P.P., Gadgil, A., & Koomey, J.G. (2002). "What can history teach us? A retrospective examination of long-term energy forecasts for the United States." *Annual Review of Energy and Environment*, 27:83–118.

Curtright, A.E., Morgan, M.G., & Keith, D.W. (2008). "Expert assessments of future photovoltaic technologies." *Environmental Science & Technology*, 42 (24):9031–8.

Dalkey, N.C. (1969). *The Delphi Method: An Experimental Study of Group Opinion*. RM-Report 5888-PR Santa Monica, Calif.: RAND Corporation. Retrieved from http://www.rand.org/content/dam/rand/pubs/research_memoranda/RM5888/RM5888.pdf (Accessed on February 12, 2014).

Davis, G.A., & Owens, B. (2003). "Optimizing the level of renewable electric R&D expenditures using real options analysis." *Energy Policy*, 31:1589–1608.

Denholm, P., & Margolis, R. (2008). *Supply Curves for Rooftop Solar PV-Generated Electricity for the United States*. National Renewable Energy Laboratory. U.S. Department of Energy Technical Report. NREL/TP-6A0–44073. November. Retrieved from www.nrel.gov/docs/fy09osti/44073.pdf (Accessed July 21, 2011).

DOE. (2008, July). "20% wind energy by 2030." Washington, D.C.: Office of Energy Efficiency and Renewable Energy. U.S. Department of Energy. Retrieved from http://www.nrel.gov/docs/fy08osti/41869.pdf (Accessed July 21, 2011).

Dooley, J.J. (2008, October). "U.S. federal investments in energy R&D: 1961–2008." Pacific Northwest National Laboratory Report PNNL-17952. Prepared for the U.S. Department of Energy under Contract DE-AC05–76RL01830.

DSIRE. (2010). "Renewable electricity production tax credit (PTC)." Database of State Incentives for Renewables & Efficiency. Retrieved from http://www.dsireusa.org/incentives/incentive.cfm?Incentive_Code=US13F (Accessed January 4, 2011).

EERE. (2008). "20% wind energy by 2030: Increasing wind energy's contribution to U.S. electricity supply." Washington, D.C.: Office of Energy Efficiency and Renewable Energy. U.S. Department of Energy.

EIA. (2009). "Energy market and economic impacts of H.R. 2454, the American Clean Energy and Security Act of 2009." Washington, D.C.: Energy Information Administration, U.S. Department of Energy.

EIA. (2010a). *Annual Energy Outlook 2010 with Projections to 2035*. Report No. DOE/EIA-0383(2010). Washington, D.C.: Energy Information Administration. U.S. Department of Energy, United States. Retrieved from http://www.eia.gov/oiaf/aeo/pdf/0383(2010).pdf (Accessed February 12, 2014).

EIA. (2010b). *Annual Energy Outlook 2011 Early Release Report*. Report No. DOE/EIA-0383ER(2011) [Homepage of Energy Information Administration. U.S. Department of Energy]. Retrieved from http://www.eia.doe.gov/forecasts/aeo/index.cfm (Accessed January 23, 2011).

EIA. (2010c). *Annual Energy Review*. Report No. DOE/EIA-0384(2009). Washington, DC: Energy Information Administration. U.S. Department of Energy.

EIA. (2010d). "Assumptions to the Annual Energy Outlook 2010." Washington, D.C.: Energy Information Administration. U.S. Department of Energy.

EIA. (2010e). Buildings Energy Data Book [Homepage of Energy Information Administration. U.S. Department of Energy]. Retrieved from buildingsdatabook.eren.doe.gov/TableOfContents.aspx (Accessed July 8, 2011).

EIA. (2011a). *Natural Gas – Data*. Washington, D.C.: U.S. Energy Information Administration. U.S. Department of Energy. Retrieved from http://www.eia.gov/dnav/ng/ng_pri_sum_dcu_nus_a.htm (Accessed July 24, 2011).

EIA. (2011b). *Petroleum & Other Liquids – Data*. Washington, D.C.: U.S. Energy Information Administration. U.S. Department of Energy. Retrieved from http://www.eia.gov/dnav/pet/PET_PRI_WCO_K_W.htm (Accessed July 24, 2011).

EIA. (2011c). *Annual Energy Review – Electricity*. Washington, D.C.: U.S. Energy Information Administration. U.S. Department of Energy. Table 8.11b. Retrieved from http://www.eia.gov/emeu/aer/elect.html (Accessed July 22, 2011).

Energy Innovation Tracker. (2013). "Energy innovation tracker." Retrieved from http://energyinnovation.us/data/ (Accessed May 2, 2013).

Fischer, C., & Newell, R.G. (2008). "Environmental and technology policies for climate mitigation." *Journal of Environmental Economics and Management*, 55:142–62.

Fishbone, L.G., & Abilock, H. (1981). "MARKAL, a linear-programming model for energy systems analysis: Technical description of the BNL version." *International Journal of Energy Research*, 5(4):353–75.

Gallagher, K.S., & Anadon, L.D. (2013). DOE Budget Authority for Energy Research, Development, & Demonstration Database. Retrieved from http://belfercenter.ksg.harvard.edu/publication/23010/ (Accessed April 20, 2013).

Gallagher, K.S., Holdren, J.P., & Sagar, A.D. (2006). "Energy-technology innovation." *Annual Review of Environmental Resources*, 31:193–237.

Garcia, M.L., & Bray, O.H. (1997). *Fundamentals of Technology Roadmapping*, SAND97–0665. Albuquerque, N.M.: Sandia National Laboratories.

Geels, F.W., Hekkert, M.P., & Jacobsson, S. (2008). "The dynamics of sustainable innovation journeys." *Technology Analysis & Strategic Management*, 20(5):521–36.

Goulder, L., & Schneider, S. (1999). "Induced technological change and the attractiveness of CO_2 emissions abatement policies." *Resource and Energy Economics*, 21:211–53.

Griliches, Z. (1992). "The search for R&D spillovers." *Scandinavian Journal of Economics*, 94:S29–S47.

Gruebler, A. (2011). "The costs of the French nuclear scale-up: A case of negative learning by doing." *Energy Policy*, 38:5174–88.

Henderson, R.M., & Newell, R.G., eds. (2011). *Accelerating Innovation in Energy: Insights from Multiple Sectors*. Chicago: University of Chicago Press.

Hicks, J.R. (1932). *The Theory of Wages*. London: Macmillan.

Holdren, J.P., & Baldwin, S.F. (2001). "The PCAST energy studies: Toward a national consensus on energy research, development, demonstration, and deployment." *Annual Review of Energy and the Environment*, 26:391–434.

Howell, K. (2010). "Sen. Graham's plan for clean-energy bill could drain RES support." *Greenwire. The New York Times*, September 29.

Iman, R.L., & Conover, W.J. (1982). "A distribution-free approach to inducing rank correlation among input variables." *Communications in Statistics B*, 11:311–34.

Jaffe, A.B., Newell, R.G., & Stavins, R.N. (2005). "A tale of two market failures: Technology and environmental policy." *Ecological Economics*, 54:164–74.

Kammen, D.F., & Nemet, G.F. (2005a). "Supplement: Estimating energy R&D investments required for climate stabilization." *Issues in Science and Technology*, 22(1): 84–8.

Kammen, D.M., & Nemet, G.F. (2005b). "Reversing the incredible shrinking U.S. energy R&D budget." *Issues in Science and Technology*, 22(1):84–8.

Laitner, J.A. (2009). The National Energy Efficiency Resource Standard as an energy productivity tool [Homepage of American Council for an Energy-Efficient Economy]. Retrieved from http://aceee.org/white-paper/national-eers-energy-productivity-tool-february-2009 (Accessed September 14, 2010).

Lester, R.K., & Hart, D.M. (2011). *Unlocking Energy Innovation*. Cambridge, Mass.: MIT Press.

Loulou, R., Goldstein, G., & Noble, K. (2004). Documentation for the MARKAL family of models. Retrieved from www.etsap.org/MrklDoc-I_StdMARKAL.pdf (Accessed February 12, 2014).

Lutsey, N., & Sperling, D. (2008). "America's bottom-up climate change mitigation policy." *Energy Policy*, 36:673–85.

Martino, J. (1987). "Using precursors as leading indicators of technological change." *Technological Forecasting and Social Change*, 32:341–60.

McKay, M.D., Conover, W.J., & Beckman, R.J. (1979). "A comparison of three methods for selecting values of input variables in the analysis of output from a computer code." *Technometrics*, 21:239–45.

McNerney, J., Trancik, J.E., & Farmer, J.D. (2011). "Historical costs of coal-fired electricity and implications for the future." *Energy Policy*, 39:3062–54.

MIT. (2006). *The Future of Geothermal Energy: Impact of Enhanced Geothermal Systems (EGS) on the United States in the 21st Century*. Report. Cambridge, Mass.: Massachusetts Institute of Technology. Retrieved from http://www1.eere.energy.gov/geothermal/pdfs/future_geo_energy.pdf (Accessed on February 12, 2014).

Morgan, M.G., & Henrion, M. (1990). *Uncertainty: A guide to Dealing with Uncertainty in Quantitative Risk and Policy Analysis*. Cambridge: Cambridge University Press.

Mowery, D.C., Nelson, R.R., & Martin, B.R. (2010). "Technology policy and global warming: Why new policy models are needed (or why putting new wind in old bottles won't work)." *Research Policy*, 39:1011–23.

Mowery, D., & Rosenberg, N. (1979). "The influence of market demand upon innovation – a critical review of some recent empirical studies." *Research Policy*, 8(2):102–53.

Narayanamurti, V., Anadon, L.D., Breetz, H., Bunn, M., Lee, H., & Mielke, E. (2011). *Transforming the Energy Economy: Options for Accelerating the Commercialization of Advanced Energy Technologies*. Report for Energy Technology Innovation Policy Research Group. Cambridge, Mass.: Belfer Center for Science and International Affairs, John F. Kennedy School of Government, Harvard University.

NCEP. (2004). "Ending the energy stalemate. A bipartisan strategy to meet America's energy challenges." Washington, D.C.: The National Commission on Energy Policy.

NCEP. (2007). *Energy Policy Recommendations to the President and the 110th Congress*. Washington, D.C.: National Commission on Energy Policy.

Nemet, G.F. (2006). "Beyond the learning curve: Factors influencing cost reductions in photovoltaics." *Energy Policy*, 34:3218–32.

NRC. (2001). *Energy Research at DOE: Was It Worth it? Energy Efficiency and Fossil Energy Research 1978 to 2000*. Board on Energy and Environmental Systems. Washington, D.C.: The National Academies Press.

NRC. (2007). *Prospective Evaluation of Applied Energy Research and Development at DOE: Phase Two*. Washington, D.C.: The National Academies Press.

Oppenheimer, M., O'Neill, B.C., Webster, M., & Agrawala, S. (2007). "The limits of consensus." *Science*, 317:1505–06.

Oster, S.M., & Quigley, J.M. (1977). "Regulatory barriers to the diffusion of innovation: Some evidence from building codes." *The Bell Journal of Economics*, 8(2): 361–77.

PCAST. (1997). *Federal Energy Research and Development for the Challenges of the Twenty-First Century*. Report. Washington, D.C.: President's Council of Advisors on Science and Technology, Executive Office of the President. Retrieved from http://www.whitehouse.gov/sites/default/files/microsites/ostp/pcast-nov2007 (Accessed on February 12, 2014).

PCAST. (2010). *Report to the President on Accelerating the Pace of Change in Energy Technologies Through an Integrated Federal Energy Policy*. Washington, D.C.: President's Council of Advisors on Science and Technology. Executive Office of the President.

Pugh, G., Clarke, L., Marlay, R., Kyle, P., Wise, M., McJeon, H., & Chan, G. (2010). "Energy R&D portfolio analysis based on climate change mitigation." *Energy Economics*, 33(4):634–43.

Ramseur, J.L. (2007). *Climate Change: Action by States to Address Greenhouse Gas Emissions*. Order Code RL33812. Washington, D.C.: Congressional Research Service. Retrieved from http://fpc.state.gov/documents/organization/80733.pdf (Accessed February 12, 2014).

Sagar, A.D. (2004). "Technology innovation and energy." Cleveland, C. (Ed). In *The Encyclopedia of Energy*, Vol. 6, edited by C. Cleveland, pp. 27–43. Elsevier Online.

Sagar, D.J., & van der Zwan, B. (2006). "Technological innovation in the energy sector: R&D, deployment and learning-by-doing." *Energy Policy*, 34:2601–8.

Schock, R., Fulkerson, W., Brown, M., San Martin, R., Greene, D., & Edmonds, J. (1999). "How much is energy research and development worth as insurance?" *Annual Review of Energy and the Environment*, 24:487–512.

Sciortino, M. (2010). State Energy Efficient Resource Standard (EERS) Fact Sheet, Washington, D.C.: American Council for an Energy-Efficient Economy.

Stewart, W.E., Stoddard, E.H., Geraci, K.K., & Ham, D.K. (1983). "Trends in federal support for energy R&D." Washington, D.C.: National Science Foundation, Division of Science Resource Studies. February.

The White House. (2009). "Press gaggle by Press Secretary, Robert Gibbs; Deputy National Security Advisor for International Economic Affairs, Mike Froman; and Assistant to the President for Energy and Climate, Carol Browner." Washington, D.C.: Office of the Press Secretary, The White House.

Tversky, A., & Kahneman, D. (1973). "Availability: A heuristic for judging frequency and probability." *Cognitive Psychology*, 5(2):207–32.

Wang, S.S. (1998). "Discussion papers already published: 'Understanding relationships using copulas' by Edward Frees and Emiliano Valdez." *North American Actuarial Journal*, 2(1):137–41.

Weiss, C., & Bonvillian, W.B. (2009). *Structuring an Energy Technology Revolution*. Cambridge, Mass.: MIT Press.

Wiser, R.H., & Pickle, S.J. (1998). "Financing investments in renewable energy: The impacts of policy design." *Renewable and Sustainable Energy Reviews*, 2:361–86.

Yang, B. (2005). *Residential Energy Code Evaluations Review and Future Directions*. Providence, R.I.: Building Code Assistance Project.

3

Reforming U.S. Energy Innovation Institutions: Maximizing the Return on Investment

NATHANIEL LOGAR, VENKATESH NARAYANAMURTI,
AND LAURA DIAZ ANADON

3.1. Introduction and Motivations

The large new investments in energy research, development, and demonstration (RD&D) recommended in the last chapter will not be justified or sustainable unless they are managed in a way that maximizes the effectiveness of the investment. Hence, the U.S. government must ensure that its energy innovation institutions are as efficient and effective as they can be.

The Bush and Obama administrations have seen substantial innovation and reform in U.S. energy innovation institutions, with the creation of the Advanced Research Project Agency-Energy (ARPA-E), the Energy Innovation Hubs, the energy frontier research centers, and strengthened connections between Basic Energy Sciences and the applied R&D offices at the Department of Energy (DOE). But as we will outline in this chapter, there is a great deal more to be done to ensure that DOE's energy innovation institutions are as effective as they can be.

DOE's national laboratories, in particular, are a core element of DOE's science and innovation system. In addition to goals of nuclear security and scientific exploration, DOE also aims to contribute to new discoveries and inventions in energy technologies. DOE itself is part of a larger innovation regime in the United States and globally that includes different actors and institutional models for conducting research. For energy innovation, a successful national laboratory outcome is often the commercialization of a technology product. Whether a laboratory's research is fundamental inquiry or it sits on the applied end of the research spectrum, it must contribute to knowledge or technologies that can eventually find use and enhance the energy system's impact on national security, the environment, and the market.

The federal science funding system has historically relied on the ideal of undirected basic research in the physical sciences (Shapley & Roy, 1985). Federally funded science has often pursued a model that disassociates the idea of useful science from the concept of basic research. None of the national laboratories focus universally on curiosity-driven fundamental research. Instead, the missions of every federal agency point them toward public benefit and social utility.

Policymakers (Brown, 1992; Marburger, 2005) and scholars (Funtowicz & Ravetz, 1993; Gibbons et al., 1994; Sarewitz & Pielke Jr., 2007) alike have questioned the processes and assumptions that drive much science and technology decision making, and have attempted to institute new frameworks for addressing the problem of funding robust, relevant research. Even with fundamental research, long considered to be optimal when scientists freely pursue ideas born of their own curiosity, many scholars have followed the lead of Stokes (Stokes, 1997) in calling for considerations of use and application. At the same time, they have asked for more thinking and new modes of managing science that would facilitate effective linkages between scientific information and public use. This chapter examines three institutions, all focused on innovation, in an effort to assess each one's capacity to contribute to successful energy technologies in light of ongoing discussions and concerns regarding the need to conduct research lending equal value to invention and discovery (Narayanamurti et al., 2013), while still allowing for a flexible and accountable research enterprise. These case studies, on the National Renewable Energy Laboratory (NREL), Semiconductor Research Corporation (SRC), and Energy Biosciences Institute (EBI), aim to make recommendations about the management of such institutions, and the higher order science policies that dictate the constraints and opportunities facing energy innovation.

National labs contribute to energy technology innovation in a number of ways: (1) national labs provided support during the Deepwater Horizon oil spill in 2010; (2) international researchers generally value the national labs and attempt to collaborate with them; (3) reports and expertise from the national labs are often used by private companies; and (4) test facilities at the labs can reduce the costs for private firms to innovate. However, there are areas where the labs lack strength, such as funding high-risk, long-term, application-inspired research (e.g., an area that ARPA-E was created to address), bridging the basic and applied research divide (e.g., an area that the Energy Innovation Hubs can contribute to), and managing large-scale demonstration projects (see the box on pp. 116–118 for the summary of a high-level workshop conducted in late 2010 about the need and possible design for an institution to promote large-scale technology demonstrations) (Narayanamurti et al., 2011). Despite this, the national labs have made significant contributions to developing and commercializing the energy technologies the United States needs (e.g., they played a key role in the development of the thin firm technology underlying First Solar, the largest U.S. photovoltaic [PV] company).[1] It is likely that some of the investments made in the labs over time will result in more benefits, particularly if DOE can remove some of the barriers to working efficiently.

[1] Unfortunately, it is difficult to measure the full extent of these contributions given that some of the labs do not keep track of their contributions in terms of technology development and transfer in sufficient detail.

More work to improve on such successes could aid DOE in fulfilling its missions at a time when energy research has become an important issue to America's continued prosperity. Most of the national laboratories sit within DOE's Office of Science (OS), which has historically focused on basic research in the physical sciences. Other laboratories work under different DOE offices, including the National Nuclear Security Administration, the Office of Energy Efficiency and Renewable Energy, the Office of Fossil Energy, and the Office of Nuclear Energy. Many analyses have identified a need for altered systems of energy innovation in DOE labs. In 1983, a report by a Federal Laboratory Review Panel, headed by David Packard, co-founder of Hewlett-Packard, advised that laboratories should work under tightly defined missions, interact closely with research users, reduce bureaucratic constraints to lab managers, and hold external reviews to make managers accountable for scientific excellence and relevance (OSTP, 1983). In an earlier article, some of the authors of this study outlined a broad set of similar recommendations for managing U.S. energy innovation institutions (Narayanamurti et al., 2009). They argue that the national laboratories originally had a clear mission, strong leadership, an entrepreneurial culture, stable funding, and a structure that allowed researchers to be independent, yet accountable and connected to the larger system. They argue that over time, the mission of the multiprogram national labs has become more diffuse and that for-profit management of some of the labs may have increased emphasis on corporate priorities, sometimes at the expense of national priorities. For too long, U.S. policymakers have focused too little attention on the performance of U.S. energy innovation institutions, including the national laboratories.

Many decision makers and scholars assume that there is a strong relationship between research funding and societal benefit. This is certainly the case when innovation institutions are managed effectively and are focused on meeting major societal needs. But in-depth independent assessments are needed to identify which new steps should be taken to improve these institutions' effectiveness. Of course, funders and other parties already undertake assessments of energy innovation institutions. Several groups regularly evaluate NREL, including technical advisory boards and decision makers in DOE. These reviews assessments certainly help, by bringing outside knowledge and perspective to the organization reviewed because the participants in such reviews rarely have the time, the resources, or the incentives needed to conduct an in-depth, comprehensive evaluation and may not always be in a position to challenge existing U.S. government approaches where necessary.

NREL and the National Energy Technology Laboratory are the two national labs in the energy area with missions that, in addition to specifying public value, guide them to conduct applied research, and NREL is the only laboratory that aims specifically at problems of renewable energy and energy efficiency. As such, it is appropriate to study the lab's record of innovation, as renewable energy and energy efficiency are growing areas of the nation's innovation concerns, with more attention being paid to problems and opportunities of renewable energy in recent years due to the issues of climate change and energy independence. With more than 1,000 employees, a yearly budget that exceeds $300 million, and large-scale user facilities for research and testing, NREL is sizeable enough to be a major part of the energy innovation apparatus, but to still work under a fairly cohesive mission. Most important, newer institutions such as ARPA-E, DOE Energy Innovation Hubs, and Energy Frontier Research Centers (EFRCs) are all working to enhance renewable energy innovation in novel ways. Each of these focuses on a different type of research, with EFRCs focusing on fundamental research on barriers to new technology development, ARPA-E on high-risk revolutionary inquiry, and Innovation Hubs on addressing energy problems in a multidisciplinary approach that addresses fundamental work through commercialization (Chu, 2009). These new actors participate in the U.S. national innovation system, where the role of the government, linked to the national laboratories, has long been criticized for being "siloed" and too fixated on maintaining existing research portfolios (Duderstadt et al., 2009). To justify itself, NREL is in a place where it must first define its role in a large, complex innovation system, and second, utilize that role to effectively encourage innovation.

In examining NREL, this chapter asks: How can those managing NREL best facilitate innovation? Factors such as mission, funding, research culture, and leadership, highlighted by Narayanamurti et al. (2009), are all important to addressing this question. Avenues of inquiry include how NREL decision makers target users, how they learn about user needs, and what formal or informal mechanisms might accomplish this. The chapter also explores what factors play a role in decisions, what mechanisms enable these things, what constraints exist, and which users might be left out.

This case, along with cases on other innovation institutions, Semiconductor Research Corporation (SRC) and Energy Biosciences Institute (EBI), will allow for a broader understanding of how energy innovation works, with a deeper insight into the operations of a few types of institutional structures. As a historically successful example of government–private–academic partnership that has led to innovation in the marketplace, we examined SRC for lessons learned from its model. SRC has consistently supported focused centers for long-term innovation work, participated in roadmapping activities for strategizing on future innovation, and developed its industry's workforce through graduate student funding. As a supporter of the semiconductors and integrated circuits industry, SRC has historically focused on funding research close to the area of semiconductors, but it is now also funding centers in Smart Grid technology and PVs. Another organization, EBI, represents a new kind of arrangement between private, university, and governmental institutions. This model fosters cooperative relationships that allow industry users, academic researchers, and funders to work on innovation in a way that benefits all parties. In addition, EBI represents an example, within energy, of an institution augmenting its existing expertise to pursue end-to-end research, including fundamental work, applied science, invention, and socioeconomic analysis of the application context. The arrangement is also an interesting long-term commitment (10 years) of sustained funding by industry for work at a university, thus providing a bridge between two modes of knowledge production.

The case studies that follow contain data about NREL, SRC, and EBI gathered in semistructured, qualitative interviews (Rubin & Rubin, 1995). Interviewees at all institutions ranged from the heads of the organizations to division leaders, laboratory staff, and those working in commercialization and analysis. In addition to interviews, much of the data came from an examination of missions, budget figures, and evaluation documents. This information provided both the specifics on how the institution operates and fine detail on the broader world of DOE national laboratories. We also reviewed records, such as external evaluations, numbers on technology transfer and patent activities, and press releases.

For NREL, 20 interviews were held with staff at the institution, along with some members of the user communities. Users included researchers and decision makers in the PV industry, along with energy lobbyists and advocacy groups. SRC and EBI interviewees included staff, sponsors, researchers, and students.

Important questions included how research decisions are made, whose input matters in decision making, how decisions incorporate the needs of users, how outcomes are evaluated, and where successes and failures for the institutions lie. The focus was on the decision processes that govern the institutions, along with how these processes feed into the accomplishment of the mission and user satisfaction. The interview process continued until everyone with responsibility for management of the institution had been heard, and we had obtained a consistent picture of institutional procedures and outcomes.

Given the increased attention that governments around the world have played to the set of institutions promoting energy innovation (Anadon, 2012), we hope that the insights from this chapter will serve to highlight the many factors that must be considered when designing these institutions and the system of institutions. This chapter discusses NREL, first by addressing its overriding mission and its context within DOE, then by examining its decision processes amid the obstacles it faces to illustrate one laboratory's attempts to reconcile its goal of innovation within its constraints. Following the study of NREL, we provide a similar treatment of SRC and EBI, and a summation of lessons that apply to effective governance of energy innovation activities.

3.2. National Renewable Energy Laboratory

3.2.1. Mission and Description

As a national laboratory, NREL works within a DOE system that includes research in energy, nuclear security, and basic science. NREL's mission supports the idea of renewable energy as a public good and seeks to "transfer knowledge and innovation" to meet energy needs, and to put new discoveries in the hands of industry users. Thus, "NREL firmly believes that it is in the nation's best interest to move technologies into the marketplace in an expeditious manner." The laboratory succeeds when it delivers useful technology outputs and outcomes to its customers.

NREL is a government-owned, contractor-operated laboratory, as are 16 of the 17 national laboratories. The Alliance for Sustainable Energy, headed by Battelle and the Midwest Research Institute, manages the laboratory under a five-year, renewable contract. NREL receives much of its funding from the Office of Energy Efficiency and Renewable Energy (EERE), with a portion coming from OS and other DOE offices. In 2010, NREL received $18.8 million of its $390.8 million, or 4.8% of the total, from OS, and $15.7 million from other parts of DOE, including the Office of Electric Delivery and Energy Reliability and other national laboratories. The laboratory director, Dan Arvizu, distributes some of this money as laboratory-directed research and development (LDRD). NREL currently sets LDRD around $8 million per year (2.6% of the lab's 2009 total) that is applied as an overhead cost to incoming funds. According to one decision maker at Sandia National Laboratories, LDRD provides a key source for "high risk, potentially high value research" (*Physics Today*, 2009).

Although many of the laboratories pursue some applied or use-inspired research, many at NREL see it as the lab that most strongly emphasizes application, as reflected in its mission. NREL's organizational divisions include Science & Technology, Commercialization & Deployment, and Outreach, Planning & Analysis. The division of Science & Technology itself consists of three laboratory programs: Renewable Electricity & End Use Systems, Renewable Fuels & Vehicle Systems, and Basic Energy Sciences (BES). The first two focus on applied programs in areas such as PVs and bioenergy. The third performs NREL's fundamental research and receives proportionally more funding from OS (Figure 3.1). OS's BES program exists "to provide the foundations for new energy technologies and to support DOE missions in energy, environment, and national security" (DOE, 2008).

As a laboratory, NREL aims at successful commercialization of its research products within the energy industry. One NREL manager stated the central concern as such:

> In the late 70s, we were developing concentrated solar power and first generation PV cells at Sandia. It was very cool from a scientific perspective, but everyone who looked at it said, "My goodness, how is that relevant in our marketplace?" And here you fast forward 35 years, that technology, developed in that era, is now today's commercial product... What do we have to do to have today's innovation not take 35 years to get to the marketplace?

Many at NREL echo the goal of commercialization evident in this statement and the mission. For example, an employee in Renewable Electricity and End Use said about his group, "the role of our center is developing energy infrastructure for commercialization."

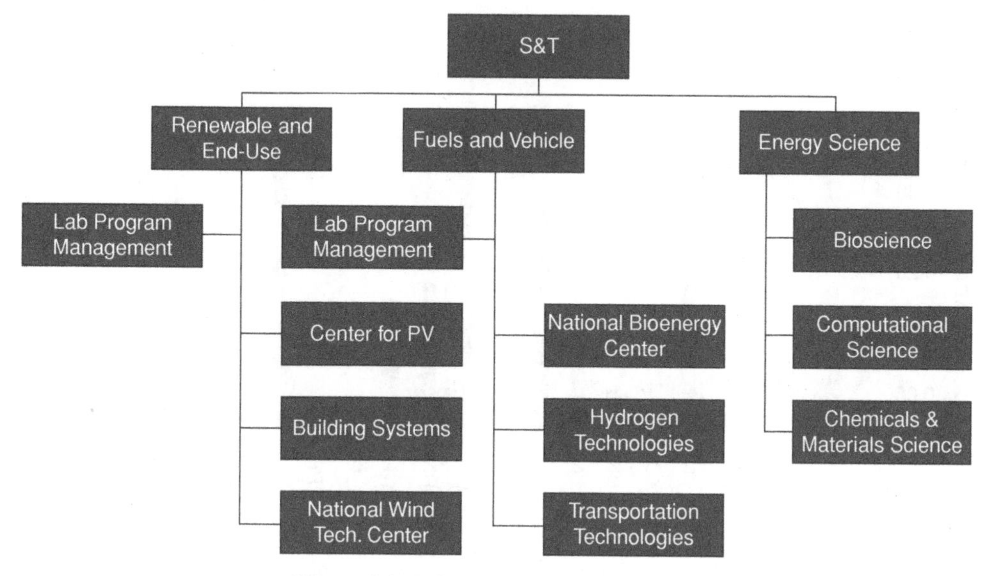

Figure 3.1. NREL organizational chart.

As NREL's manager, the Alliance for Sustainable Energy sees its role as improving the impact NREL has on industry. All members of the management team stressed this part of NREL's role, and the latest contract was altered to emphasize commercialization. For this reason, NREL instituted the division of Commercialization & Deployment. With leadership at the laboratory vice president level, this division is meant to ensure that commercialization concerns become a part of high-level decisions, integrated with other laboratory activities.

3.2.2. NREL Performance

The laboratory has had successes in promoting commercialization and in supporting the interests of industry. In 2009, NREL had 144 active Cooperative Research and Development Agreements (CRADAs), bringing in $4.9 million from industry. Sixty-seven of these CRADAs were with small businesses; 50 of them were new in 2009. In addition, 115 Work for Others projects accrued $10.6 million, 9 U.S. patents were issued, and the lab had 101 active licenses in that year. NREL also had 34 new licenses and 25 patents (worldwide) issued in 2009. In most of these categories, the laboratory is on the higher end of DOE labs, once budget size is accounted for, and is first in number of CRADAs.

Figure 3.2 shows the number of CRADAS for the year 2009 across national laboratories, along with the number of CRADAs per million budget dollars. The high numbers for NREL relative to other labs might represent one area where adherence to its mission is evident. The laboratory received almost $5 million from CRADAs in

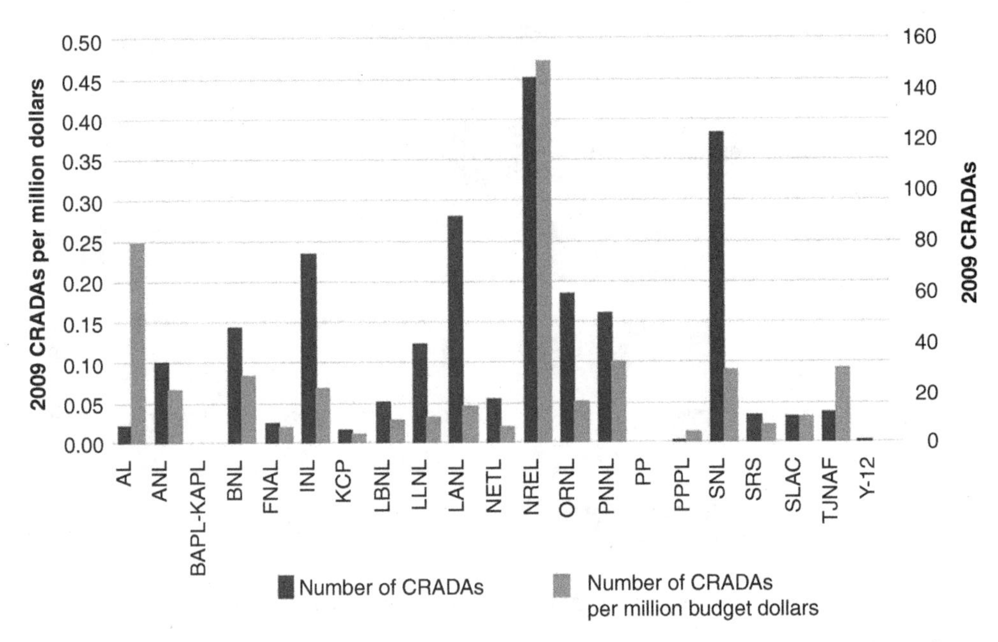

Figure 3.2. 2009 CRADAS across national laboratories, along with number of CRADAs per million budget dollars.

that year. For licenses, NREL has one of the lower numbers when compared to other laboratories, after accounting for budget, but was in the top quartile in terms of new licenses for 2009 (Figure 3.3). The lab was also second in patents issued per dollar (Figure 3.4), and ranks highly in non-federally sponsored Work for Others project, both in terms of number of projects and amount of funding (Figure 3.5).

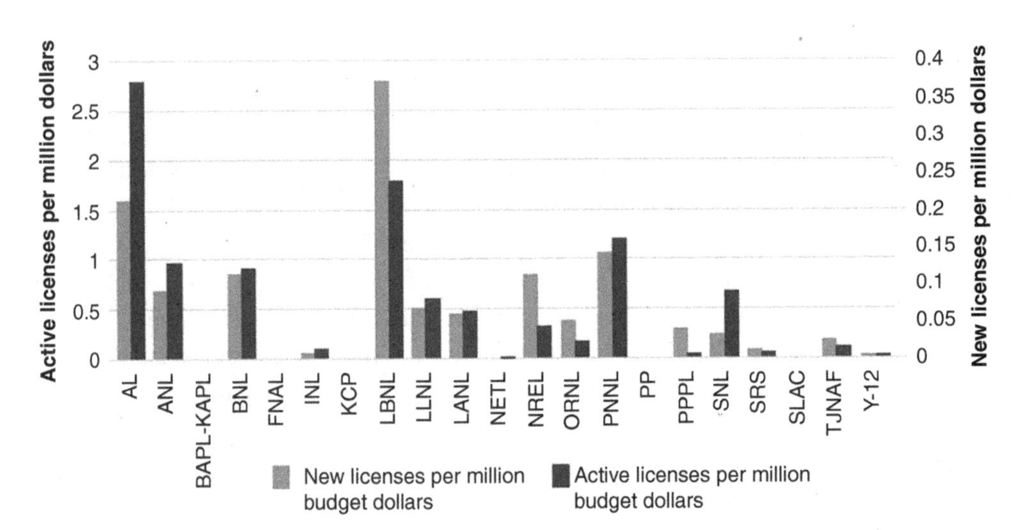

Figure 3.3. Active licenses and new licenses in 2009 per million budget dollars, across laboratories.

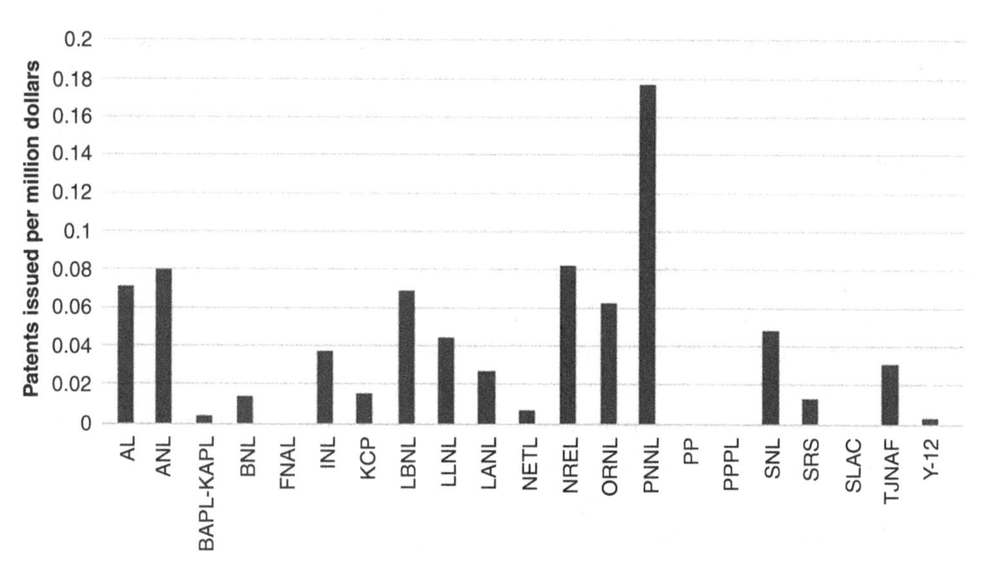

Figure 3.4. Patents issued per million dollars in 2009, across laboratories.

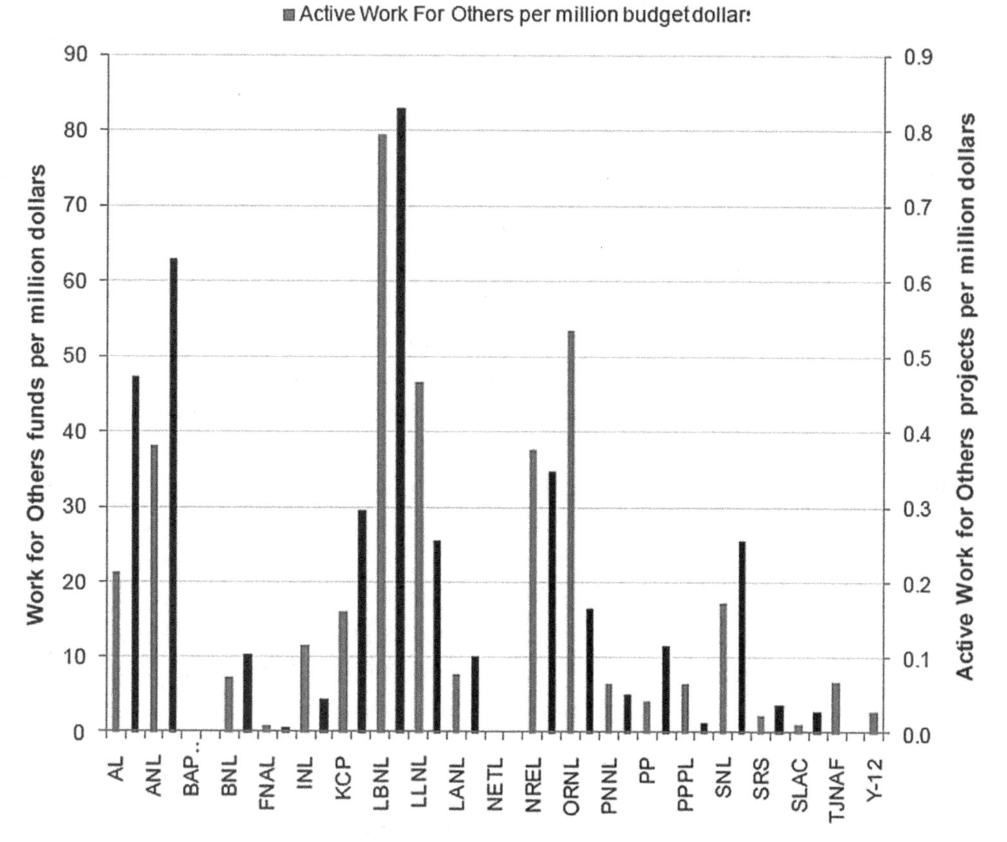

Figure 3.5. 2009 Funding in from Work for Others (from non-federal sponsors), along with the number of active Work for Others agreements, divided by operational budget.

Staff also had many stories of successful relationships with industry, particularly in the solar sector. PV companies, especially those using thin film technologies, have benefitted from relationships with NREL. First Solar, now the country's largest thin film manufacturer, is one such case. NREL began working with people from the company in 1991, when it was still named Solar Cells Inc. Many interviewees characterized the relationship as extremely close, with NREL working hand in hand with the company during its early years. Together, NREL and First Solar share an R&D 100 Award for development of a high-rate vapor transport deposition process for high-volume manufacturing (First Solar, 2003).

NREL has fostered several successful relationships through its testing activities and user facilities. It has worked with companies such as Southwest Windpower and Clipper Windpower to validate designs and field-test new turbines (EERE, 2006a). NREL also has the testing capabilities and facilities for user-led research. NREL makes many of its laboratories available for industry, or other groups, to perform or benefit from research and development. A few of the many facilities are an Advanced Research Turbines Facility, an Integrated Biorefinery Research Facility, an Outdoor Test Facility for PVs, and Wind Turbine Test Pads. Such facilities allow users to coordinate their own technology products with NREL work and facilitate formal, organized involvement by industry in renewable energy research.

3.2.2.1. Budgetary Constraints

NREL also has many obstacles to fulfilling its mission. The first among these is the laboratory's budget. NREL began as the Solar Energy Research Institute under President Carter, but solar energy began to see reduced support during the end of his presidency (Laird, 2001). Reagan withdrew support further; NREL's budget was cut in 1981 from $135 million to $100 million (a cut from $325 to $239 million in CPI-adjusted 2010 dollars), and reached a nadir of $25 million in the mid-1980s. Though the laboratory received more support under Clinton, it was not consistent. In 1996 the laboratory was faced with a 10% workforce reduction due to "lower than expected appropriations for renewable energy programs" (NREL, 1995). In 2006, after George W. Bush used his State of the Union address to proclaim the U.S. to be "addicted to oil," NREL faced a shortfall of $28 million, due to earmarking, and had to lay off 32 staff (EERE, 2006b). NREL eventually was able to offer many of the staff their jobs back. Under the American Recovery and Reinvestment Act, NREL received $110 million, much of which went toward buildings (NREL, 2009). This infusion is beneficial for the lab; however, it is no guarantee that NREL will not again suffer shortfalls and variations in support. See Figure 3.6 for the NREL budget variations in the past decade.

Such budget variability impedes NREL's ability to conduct its research. For an institution to do end-to-end research that includes fundamental research, applied

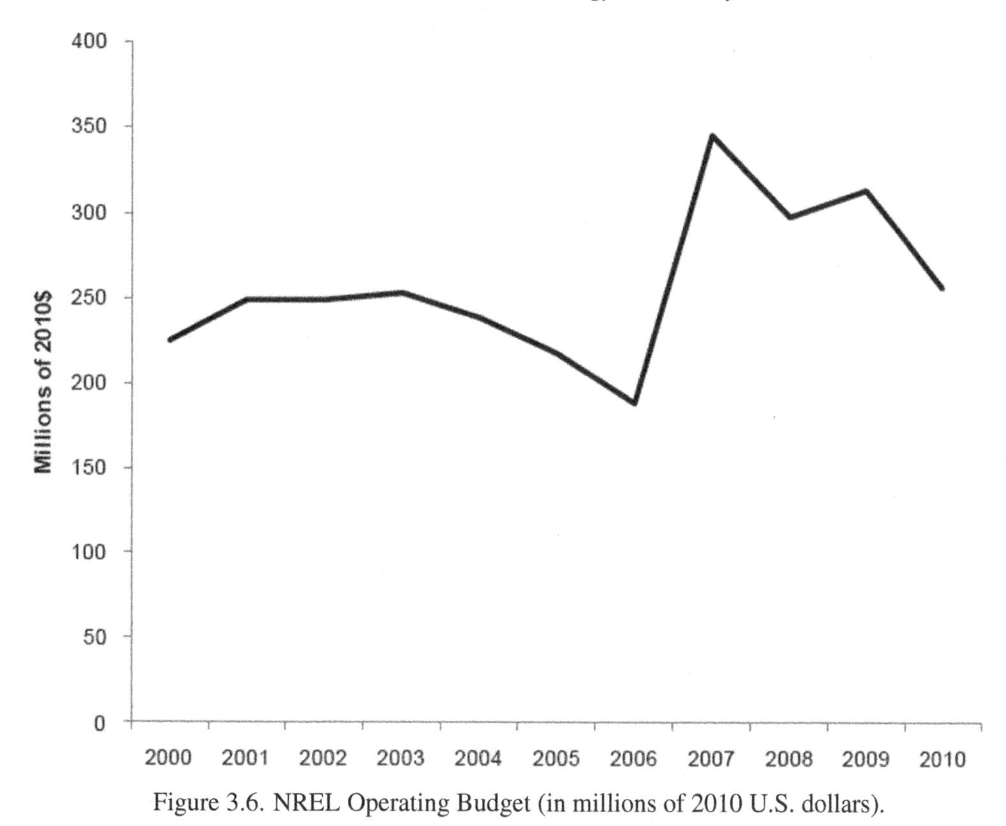

Figure 3.6. NREL Operating Budget (in millions of 2010 U.S. dollars).

science, and technology development, it requires commitment on a multiyear scale that will allow it to develop and pursue promising projects and programs, especially given that the time from concept to market for some solar technologies has been more than 30 years. While research programs should be closely evaluated and phased out if they are not working or have succeeded so the that the private sector can carry the technology forward, the effectiveness of the labs will improve with more stable funding.

3.2.2.2. Energy Markets and Policy

According to the laboratory's management, NREL's biggest problem involves navigating the complicated context posed by energy markets and public policy. For example, the year-to-year differences in wind power development closely follow trends in the production tax credit (PTC), where the individual years in which Congress has allowed the PTC to lapse correspond to drastic lows in development of wind power capacity (Wiser, 2007), highlighting the influence of factors outside of the realm of R&D on the commercialization of renewables.

In addition, the fluctuation in the price of competing technologies, such as power from natural gas, makes pegging an innovation to a specific price difficult. Obtaining a lower price per kilowatt-hour for a technology does not necessarily result in deployment, given the lack of long-term certainty about the price of competitors, along with the infrastructural, behavioral, and institutional hurdles to adoption.

3.2.2.3. Administration and Management

NREL also faces factors within DOE that shape its performance. One of these is related to the laboratory's status as a government-owned, contractor-operated lab. NREL's operating contract goes to bid every five years, and though periodic renewal of contracts is necessary to ensure the vendor is working satisfactorily, it can also put the lab in a situation where it is constantly reorganizing itself. These changes are pursued for the purpose of increasing effectiveness, but given the constant change in the renewable energy landscape, what is seen as effective during one contract renegotiation may be out of fashion five years later. While the vagaries of budget, market factors, and regulation add to this uncertainty, contract bidding encourages the institution to reorganize, and perhaps become more complex, in an effort to improve (or "reinvent") previous operating contracts and management structures.

Another issue related to administration is that of procurement. Contracts used to be handled at NREL, but they have moved under the purview of EERE. Several people at NREL opined that this reduced NREL's ability to make effective decisions, as their staff has a uniquely deep knowledge of the fields in which they work. It also impedes their ability to produce social capital, owing to the reduced interactions between NREL employees and users.

Although some decisions about budgets and priorities have to come from within DOE, NREL's staff has both technical qualifications, and relationships with people in industry that can aid them in contributing to decisions about grants, cooperative agreements, and other procurement contracts. A 1983 Packard report concluded that the national laboratory innovation system would improve if more budgetary control were given to laboratory directors (Packard, 1983; OSTP, 1983) while the trend has been for less control.

3.2.2.4. Basic Research and Applied Research

Part of a manager's role is to bring all of the parts of an organization together. Fundamental research should be insulated from the applied research concerns such as short timelines and immediate commercialization. At the same time, decision makers cannot isolate research from considerations of eventual application, and the results of basic research must be eventually integrated into applied research programs to become worth the investment. NREL faces obstacles in encouraging research to be focused toward the achievement of mission goals. For the basic research at the laboratory to succeed, it should, at the minimum, consider potential applications and how the work relates to NREL's goals. Ideally, successful basic research projects would be conceptualized so that they could be picked up by applied research and technology development projects that work toward a deployable output.

Several interviewed NREL staffers conceptualized basic and applied research as inextricable parts of a larger process that works toward application and innovation. Evaluators of NREL have also supported working to connect fundamental research to application (EERE, 2006b). Basic research at NREL is typically defined as being "use-inspired," in the phrasing of Stokes, who described such research as pursuing fundamental questions with an eye toward eventual uses (Stokes, 1997). However, most of the links between fundamental and applied projects at NREL are informal. Informal linkages can still result in coordination between basic and applied research, but they do not guarantee a sustained relationship between the two. For example, time constraints could impede staff from building such links if they are not explicitly part of the process. Responsibility for brokering such linkages can always be assumed to belong to someone else. In addition, real resources need to be committed to building such programs, either by paying for the time of people designing basic research for application, or committing funding to perform the applied research program that might continue to build on the promising findings of successful basic research, building linkages requires commitments of funding and time.

If basic research is not connected to application in some way, there is the danger that it becomes isolated from activities that are more intimately connected to users, and caught on the wrong side of the "valley of death." The U.S. Global Change Research Program neglected to consider application when promoting its basic research agenda, and as a result conducted ample high-quality science that failed to address its stated goals (Pielke Jr., 1995; Pielke Jr. & Byerly Jr., 1998; Pielke Jr. & Sarewitz, 2003). From the organizational chart of NREL (Figure 3.1), one can see that there is isolation between basic energy science and applied research, because the Energy Science division's work is further from application than that in Renewables and End-Use or in Fuels and Vehicle. Energy Sciences at NREL receives funding both from EERE and OS, but it gets more money from OS than other parts of the laboratory do.

While OS funds BES to support research that is both fundamental and foundational, EERE funding aims at "[b]ringing clean, reliable and affordable energy technologies to the marketplace." For the OS-supported research to truly serve as a foundation, active effort must be put into linking it to applied work. Doing so can be difficult because there is less industry involvement in such areas, as they are "not directly invested" in the outcome:

> Industry is interested in, "Can I fabricate this? Can I make a low cost solar cell?" The answer's yes, but they're not working on the transport and recombination mechanisms. [They would say] "Once you've got that figured out, give me something we can use." *(NREL management team member)*

One member of NREL's management team recognized the obstacles to connecting basic and applied research. Although this awareness is beneficial for the laboratory in future attempts that work to link the two, it also means NREL has not fully achieved integration of application and technology considerations into the basic research process.

3.2.2.5. User Interaction

Interaction with user groups can be important for assessing their needs, for keeping them apprised of research results and potentials, and for building social capital between institutions and stakeholders. According to many of the NREL interviewees, as well as several of the industry interviewees, NREL does have a deep understanding of user needs within their field.

Criticisms by other parties, however, show a perception that the institution falls short in providing access to industry (e.g., Serface, 2009), or that specific subsectors of an industry receive attention according to NREL's expertise or EERE considerations. At the same time, there are many examples of companies working cooperatively with NREL. This work can range from testing of technologies, to Work for Others agreements (such as Technology Services Agreements, where companies pay NREL researchers to contribute their time and knowledge) to CRADAs. The Stephenson–Wydler Technology Act of 1980 (PL 96–480) instituted CRADAs as a tool for speeding commercialization, and they serve as a means for the laboratory to share time and materials with partners in pursuit of common goals. NREL is very active in these types of agreements (Figure 3.2). When the numbers are normalized for budget amount, NREL has two times as many active CRADAs as any other laboratory, and ranks first in total money received from its CRADAs, earning a high amount of funding from non-federal sponsors.

NREL has also utilized EERE's Entrepreneur in Residence Program (EIR) to encourage interaction and commercialization (NREL, 2010; EERE, 2011). EIRs are a pilot program that puts venture capitalists from firms in labs for 18 months, where they are supposed to learn about the technologies and transition some into profitable innovations. EIRs existed in Silicon Valley firms before the national labs began using them. The EERE program, however, has not had positive results to date. One EIR, at Oak Ridge National Laboratory, said, "It was fascinating but also very disappointing. I looked at 1500 invention disclosures and walked away from 1500 ideas" (Vance & Miller, 2010). Michael Bauer, the entrepreneur, attributed the failure to a difference between the information technology field and the energy arena, where failures of commercialization are tolerated more in Silicon Valley. Joel Serface, who was EIR at NREL, thought timing was the problem:

> In the 11 months that I had the privilege to work inside NREL, I met with more than 300 researchers, identified around 30 promising technologies that I thought could reach

commercial potential over the next several years, and honed [*sic*] in on three technologies that showed imminent promise. Unfortunately, the EIR program was timed too short to reach its full potential and to get the first one of these ideas set up as a company. (Serface, 2009)

Serface also regarded the experience as one that probably aided NREL and DOE:

DOE's calculus was that if they inserted a serial entrepreneur/investor backed by a brand-named VC firm into a lab that magic would happen and that an innovation would turn immediately into a company. At worst, DOE would learn a lot about what it and its labs need to do better to in order to accelerate ideas to market.

The EIR program at NREL as well as other labs may not have led directly to commercialization, but the ancillary benefits, in the form of social capital and knowledge of the market context, could aid labs in future attempts.

Another aspect of NREL that could be improved is that the lab does not facilitate having lab scientists work for periods of time longer than six months (for one or two years) in the private sector or other institutions that may enrich their market perspective or provide complementary skills. Other labs, such as the Sandia National Lab, have managed to create this flexibility that contributes to increasing the connectivity between lab scientists and private firms.

3.2.2.6. Commercialization and Deployment

NREL, and its funders at EERE, have tried to stress increased commercialization as a means to enhance the impact of laboratory activities. Through licensing innovation to industry, NREL can facilitate the adoption of new renewable discoveries. NREL often licenses nonexclusively to those who want to utilize its technologies. For example, the institution licensed copper indium gallium selenide technologies six to eight times to different industry concerns.

Normalization of cross-laboratory data based on funding level shows that, with respect to active licenses (Figure 3.3, 2009 data) NREL had a very low number compared to other laboratories. However, when taking into account its budget, it is above the mean across national labs for new licenses and patents issued. One NREL employee in commercialization claims that NREL's overriding focus on building industries led to the success of companies such as First Solar, but that NREL did not pay attention to garnering licensing fees because it wanted the industry to easily acquire and use the results of laboratory research. First Solar is seen as a successful case because NREL played a role in First Solar's early development and was able to transfer technologies to the company without high prices.

Continued focus on commercialization, and deployment can help NREL to move new innovations into the marketplace, to maintain relationships with involved industries, and to provide context for the scientific research NREL conducts. Another means for providing such context and guidance is evident in the laboratory's Strategic Energy Analysis (SEA) division.

3.2.2.7. Strategic Energy Analysis

The SEA division of NREL encourages systems integration and attempts to "increase the understanding of the current and future interactions and roles of energy policies, markets, resources, technologies, environmental impacts, and infrastructure." Many of the NREL management regard SEA as a means for the institution to remain relevant. First, SEA allows the laboratory to conduct relevant social science and policy assessment. Second, increased knowledge of the sociotechnical space allows employees to understand the context of their innovations, and enables more informed strategic planning activities. Thus, SEA can serve as a means for integrating considerations of use into future NREL products.

SEA also has the potential to aid in integrating NREL strategy, bench level research, industry needs, and higher-level concerns. If researchers can learn about the needs and expectations of groups in industry, and the technical possibilities for research, and inform that with its knowledge of regulatory, policy, and economic issues, SEA may be able both to provide input to research prioritization and planning decisions, and to form needed connections between different groups. Recently, NREL has worked to aggressively grow the strategic energy analysis component of NREL, moving from a few employees to more than 100 in the division.

SEA alone does not guarantee that NREL's work has relevance, completes its mission, or performs transformative research in renewable energy. However, it does represent a step toward placing NREL within the larger context of the energy innovation system. NREL faces constraints, both external and internal, but it can also work within these constraints to adapt its processes to changing energy contexts and new innovation regimes. NREL's constraints include those mentioned above, such as budget, ability to apply technical knowledge to decisions, and the ability to conduct both basic and applied research that work toward industry needs. However, the laboratory is attempting to apply its technical abilities to solving pressing renewable energy innovation problems in a way that includes the user communities and transfers innovation into their hands. Progress in these areas may not be complete, and NREL and EERE might have to work further to continue such work in the face of changing policy regimes, budgets, and governmental priorities. Part of ensuring that the lab can be effective in the future involves assessing the activities of other institutions, and identifying what they do to improve the relevance of their work.

3.3. Semiconductor Research Corporation

One strategy to inform future directions at NREL, the other national laboratories, and energy research in general is to examine other institutions, organizations, and sectors. For example, an examination of SRC, an industry consortium that has enjoyed sustained support from its members and is now beginning to work on energy problems, brings out several recommendations for institutional management that can apply to the question of managing NREL and the broader issue of managing national labs.

3.3.1. Mission and Description

Robert Noyce, a co-founder of Intel and co-creator of the integrated circuit, announced the creation of SRC in 1981 as an endeavor by the Semiconductor Industry Association. A group of paying industry members funded SRC to invest in relevant academic research, intending it to help the United States compete in the international semiconductor market. Over time, SRC has expanded from its original research program, now called the Global Research Collaboration (GRC), to include two additional major initiatives also focused on semiconductor-related research, called the Focus Center Research Program (FCRP) and the Nanoelectronics Research Initiative (NRI). SRC also established two newer topical research programs, the National Institute for Nanoengineering and the Energy Research Initiative (ERI). Each program is a legally distinct entity that is a wholly owned subsidiary of SRC. The programs address different time frames and technology areas, and thus also differ somewhat in industry membership, structure, and management.

The mission of SRC is specific and clear: It is "to manage a range of consortial, university research programs some of which are worldwide, each matching the needs of their sponsoring entities. SRC maximizes synergy between programs to optimally address members' research needs and minimizes redundancy to maximize value to common members" (SRC, 2011a). SRC works toward these goals by working at the boundary between the industry members who pay into SRC and the university academics that SRC funds to perform research. SRC reinforces the clarity of its mission with abundant involvement of the industry members. Central to the achievement of the mission is graduate training in semiconductor research and other efforts to continue semiconductor innovation in the medium and long terms.

As a type of industry consortium, the institution is not subjected to many of the management issues that institutions within the government face. For example, the funders are from the same organizations as the customers, so that SRC is forced to meet user needs to retain its funding. Because the member companies fund the work, serve on SRC evaluation boards, and interact with the academics who perform the research, the system guarantees that a company can be actively involved in working to get what it wants from the process. In addition, the companies have the ability to leave SRC if they are not pleased with its direction and results, providing incentive for SRC to keep its members satisfied.

Finally, where all funds from companies used to go into one common pool, SRC has changed policy in recent years to allow companies to direct a portion of their funding to projects that they believe are promising. This allows members to continue leveraging the larger pool of funding that SRC provides, as they are not the only group contributing to a particular research thrust. They are directing a portion of their funds to buttress further a research area that they would like to support. SRC has shown itself to be flexible in its policymaking as well, with the ability to alter other processes to meet member companies' needs evident in many of its decisions. Because SRC requires the continued approval and participation of members, it is forced to be entrepreneurial. All of these characteristics point to SRC as an institution that has a mission that its member companies buy into and one that continues to produce results that these companies believe are pertinent. This institution is not useful for every semiconductor corporation; many in the field do not buy in. Those that abstain often do so because they do not see SRC research as useful to them. For example, relatively few companies work in semiconductor fabrication, which some SRC research is more pertinent to, although SRC does prioritize design work and other aspects of the industry as well. Many start-ups do not join because of their relatively small size. Though such companies might also gain in social capital and in relationships with other businesses from working together in a consortium, and though they might also see workforce benefits from student development, they do not see these as worth the price of admission due to their typically smaller budgets.

Added to this clarity of mission is the simplicity in process that allows SRC to operate cleanly and efficiently. The member companies pay money according to set rules that consider the size of the company. Once SRC has paid its operating cost, it distributes the money per the recommendations of advisory boards sent from the same industry members. Once projects are chosen based on their desirability to industry, both SRC and member companies interact with the researchers at annual reviews, conferences, and liaison interactions. The number of actors is relatively few, as is the number of transactions. And, though oversight is present, the amount of guidance/steering that SRC boards conduct during research implementation is limited.

3.3.2. SRC Programs

SRC used to consist of a single program, but has added several more over the years. Each program has its own set of member companies that buy into the mission of that program (Table 3.1). Though many companies buy into more than one program, participation is contingent on the corporations' interest and willingness to become involved in each program individually.

The Global Research Collaboration (GRC) is the part of SRC that encompasses the original program. GRC exists to fund relevant research that can help to extend the capabilities of complementary metal-oxide semiconductors (CMOS). The purpose is to keep stride with Moore's Law, which describes the trend of doubling the number of transistors on integrated circuits every two years, and thus increase attributes such as

Table 3.1. *SRC research programs*

	Global Research Collaboration (GRC)	Focus Center Research Program (FCRP)	Nanoelectronics Research Initiative (NRI)	Energy Research Initiative (ERI)
Nature of research	Narrowing options – individual researchers and centers	Creating options – multi-university centers	Discover next-logic switch – individual researchers (at NSF centers) and multi-university centers	Approaches to electrical energy generation, storage, distribution, and infrastructure
Technology spectrum	CMOS	CMOS + “hooks” to beyond CMOS	Beyond CMOS	Enabling technologies in PV and smart grid
Industry involvement	Strategic and in execution	High-level strategic	Influence through highly leveraged co-investment	Strategic and in execution
Government involvement	Limited; leverage funding	50:50 with DoD; program direction	Major funding – state and federal (NSF and NIST)	None, but SRC is open to the possibility
University autonomy	Collaborative (industry/ university)	Significant	Significant	Collaborative (industry/ university)
Membership	Global	United States only	United States only	Global
Technology transfer	Industry-driven processes (e.g., industrial liaisons, e-workshops, etc.)	Primarily faculty-driven processes (e.g., periodic reviews, seminars, etc.)	Combination of industry and faculty driven processes (e.g., industrial assignees, e-workshops, etc.)	Industry-driven processes (e.g., industrial liaisons, e-workshops, etc.)

Adapted from SRC figure.

memory and processing speed. GRC typically operates through three-year contracts between SRC and the university researchers. In addition to the money that industry contributes, GRC leverages significant funding, often from government sources, to contribute to the research. Between 1997 and 2010 the amount of leveraged funding – including collaborative funding, SRC-directed money, and SRC-influenced funding – varied between 8% and 73% of total funding. It's been greater than 60% for the last 6 years. The overhead costs for GRC are typically 12% to 13% of the funding provided by industry.

Because the member companies ultimately manage GRC, the Board of Directors consists of representatives from the companies and has overall responsibility for GRC's direction. In addition, there is an Executive Technical Advisory Board to direct the program. In the five science areas within GRC, a Science Advisory Coordinating Committee participates, and a Technical Advisory Committee helps to direct each thrust. Science committees include Device Sciences, Computer-Aided Design and Test Sciences, Integrated Circuits and Systems, Nanomanufacturing Sciences, and Interconnect and Packaging Sciences. Thrusts are topical subdivisions within these science areas, and can change depending on where the member companies believe significant academic developments can help them.

The Focus Center Research Program (FCRP) looks to work on technological changes that are farther out than the 3- to 5-year span addressed by GRC. In doing so, FCRP works by funding interdisciplinary centers that are typically led by researchers at one university, but integrate multi-university teams for solving problems. In addition to its member companies, FCRP receives approximately half of its funding from Defense Advanced Research Projects Agency (DARPA).

NRI is another program under the SRC umbrella. It exists to look beyond the incremental changes that comprise fulfilling Moore's Law, and to harness open industry participation in next generation technologies (Apte & Scalise, 2009). In addition to its corporate members, NRI has the participation of the National Institute of Standards & Technology, and strategic partnerships with the National Science Foundation, the Semiconductor Industry Association, and four state governments.

SRC's most recent program is the Energy Research Initiative (ERI), which exists to leverage industry member funding into academic discovery in precompetitive energy technologies. The mission states that it will "enable reliable, low-cost renewable energy systems and efficient energy use and distribution through an enabled and optimized smart grid" (SRC, 2011b). ERI currently focuses in two places: Smart Grid technology and PVs. While the program is just beginning, ERI provides insight into how SRC might extend its model to be influential in new fields. Members of ERI include some companies that participate in other SRC programs, such as IBM and Applied Materials, and some companies that are closely tied to the energy industry, such as thin film manufacturer First Solar, power technology company ABB, and cabling system manufacturer Nexans.

3.3.2.1. Leadership and Advice

SRC's leadership structure is fairly simple, allowing for efficient management within the corporation along with participation by industry members through a number of advisory boards. Many of the leaders of SRC have been involved with the institution for a long time, and its president, Larry Sumney, has been running the institutions since its inception in 1982. This 30-year consistency in the face of changing market dynamics and industry needs speaks to the ability of SRC leadership to continue delivering results to its members. Much of the role of SRC leadership is to serve as a middleman or liaison between the interests of industry and the academic researchers it funds. As such, the institution's advisory boards play a large role. These advisory boards consist of representatives from the member companies, and direct both the short-term technical direction and the strategic planning of SRC.

Interactions during the prioritization process and during research itself drive SRC. According to one author, "Inputs from SRC member companies through the board of Directors, the Technical Advisory Board and the mentors assures the continued responsiveness of SRC research to industry technology needs. The challenge is to distill from the variety of inputs, a coherent agenda that is consistent with both the needs of a majority of the membership" (Burger, 2000). One of the ways that industry advisors have inputs is by rating projects that SRC is considering or conducting. Members can rate projects using an SI (Success and Impact) score, which stands for a number that is the product of ratings for both the project's impact and chances of success. By integrating such scores into the decision process, SRC can alter its path toward those ideas that users feel are more likely to contribute benefit, and can thus be flexible to keep their members satisfied.

3.3.2.2. Other Processes

In addition to broad guidance by the advisory boards, many of the projects in SRC programs receive guidance from yearly program reviews. These reviews serve as a means for researchers to disseminate their results within SRC, and for members to critique the work and steer the direction for the next year. While projects usually fulfill their three-year terms, SRC programs can use annual reviews to guide and shape them as they develop.

A more personal, iterative relationship is also encouraged by the existence of industrial liaisons. Liaisons, who are employees from the members, work with the university researchers to provide input on a project so that it has a better chance of meeting an industry need and of eventually being commercialized. By making regular personal contact with the researchers, liaisons can help both the mentoring process for graduate students and the likelihood of a usable outcome.

Finally, SRC operates Techcon, a research conference and job fair that allows people working on SRC research to present to, and interact with, people in the industries that are paying them. According to interviewees who were past SRC-funded graduate students, Techcon was both a useful venue for learning about the state of the research and industry, and for securing the jobs they worked in following their studentships.

3.3.3. SRC Performance

In 2005, SRC won the *National Medal of Technology*:

> For building the world's largest and most successful university research force to support the rapid growth and advance of the semiconductor industry; for proving the concept of collaborative research as the first high-tech research consortium; and for creating the concept and methodology that evolved into the International Technology Roadmap for Semiconductors. (U.S. Patent and Trademark Office, 2005)

Several large companies have consistently funded SRC for more than 30 years. The willingness of companies such as IBM, Intel, and Texas Instruments to stay involved with SRC is evidence that the organization has been consistently meeting the expectations that its member companies set for it. This ability to keep its constituents satisfied manifests itself in several outcomes.

In terms of results, SRC has been successful in three ways. First, the research has helped to push technology forward to keep pace with Moore's Law and to contribute to other developments that will move innovation. Though this first success does not solely aid the member companies, it does help them meet emerging market needs. The National Medal of Technology credits SRC with supporting growth in the industry. In 2009, 12 patents were issued for GRC research and 12 for FCRP. This is far from the only way success in technology can be measured, and it does neglect all non-market benefits, such as workforce development, capacity building, strategic orientations, and knowledge transferred through interactions between researchers and users.

Second, SRC has continually educated graduate students in semiconductor science and engineering, thereby providing a workforce for its member companies. When the SRC began, the vast majority of graduate student research was aimed at materials other than silicon. SRC's strategy was to inject academic funding into the field in which they operate. By ensuring a significant number of emerging PhDs with research skills in pertinent areas, SRC has created a valuable commodity. Building human capital, both through the relationships created between industry and academic experts and by actually developing a workforce fluent in semiconductors, is an important part of technology transfer that has been often overlooked in trying to measure the effectiveness of technology activities (Bozeman, 2000).

Finally, SRC has continuously contributed to road mapping activities for the semiconductor industry, which has been a successful way not only to strategize the future direction for SRC and its companies specifically, but also for the larger semiconductor area within which they operate. SRC utilizes such roadmaps to deduce the direction of its thrusts and science areas.

3.4. Energy Biosciences Institute

3.4.1. Mission and Description

BP's contracts with EBI members began in 2007, when it committed $500 million dollars for a 10-year term (EBI, 2007). The University of California- Berkeley (UCB) responded to BP's call for proposals, then reached out to the University of Illinois Urbana-Champaign (UIUC) as a partner before contracting was finished, to complement their expertise, especially in the area of crop sciences. The contract stipulated that $35 million per year would go to an open research component, including work done at UCB, UIUC, and Lawrence Berkeley National Laboratory (LBNL). An additional $15 million per year was provided for a proprietary research component undertaken by offsite BP contractors, for which the goal was to aid in transitioning EBI discoveries to BP deployment Figure 3.7). The EBI contract mandates basic and applied research "to explore the application of modern knowledge to the energy sector."

EBI funds six research themes: feedstock development, biomass depolymerization, fossil fuel bioprocessing, carbon sequestration, socioeconomic systems, and biofuels production. About 80% of the institute's effort addresses the development of second-generation biofuels. The remainder of funding goes to other applications, such as using biological agents to process fossil fuels. EBI funds two research mechanisms, called programs and projects. Programs are larger in scale, longer term, and include more participants; they are broader efforts aiming at the main questions EBI is trying to answer. Projects are smaller in scale and ask more focused questions. Though a governance board holds the responsibility for top-level leadership at EBI on a long-term basis, many of the decisions on research activities and budgeting are made by the EBI management team, including EBI Director, UCB professor Chris Somerville, along with its Deputy Director, Stephen Long of UIUC, Associate Director Paul Willems from BP, and a couple of other employees. The EBI executive committee also makes decisions on budgeting and research prioritization. The executive leadership includes the management team along with representatives from each institution and from each EBI research area.

3.4.2. Research Management, Operations, and Culture

EBI aims to develop both innovation and knowledge to move past existing bottlenecks impeding the successful development of cost-effective biofuels. Chris Somerville claims that EBI work may direct inquiry into specific technologies to address such bottlenecks, but also invest in understanding the systemic context of application, including devoting money to socieconomic research in areas related to biofuels development, such as biofuels effects on crop pricing. According to both the management team and members of the executive leadership, BP contributes to EBI decision making, but not to the extent that EBI must pursue work only in BP's favored research areas. Several interviewee maintained EBI pursues its course of research with minimal interference from BP. Although the company's representatives met some early EBI programs with skepticism, EBI leadership has sought a broad, diverse portfolio, and has since convinced BP of the value of such. According to one EBI presentation (Somerville, 2011), the active management of budget and research priorities by a small team allows for a more flexible and dynamic budgetary environment.

The EBI research prioritization and execution strategy allows problem definition and scoping by the UCB, UIUC, and LBNL Principal Investigators (PIs) within the institute. The BP members of EBI are involved, and BP as a greater entity has input in the post-proposal stage, but much of the conceptualization occurs within EBI itself. BP employees, working both at the director-level and as technical advisors, work within EBI to help maintain awareness of eventual use. This allows for rapid response on new ideas; continual interaction fosters a system where the PIs' experiments can lead to inventions that BP may opt to claim as IP and commercialize within the company.

While EBI constantly monitors its projects and programs, an institute-wide formal review period is held twice a year, in both UCB and UIUC. One result of such reviews is program budgeting for upcoming activities. For the management team of EBI, this can include very detailed discussion about their PIs' goals relative to our view of the EBI portfolio. An additional task involves cutting funds from projects and programs that are not likely to be productive, or are not meeting EBI's mission. Below is one example of those decisions, and how EBI approaches them:

From the beginning we had several of the chemical engineering faculty, who believed that an interesting approach to the overall process would be to use something called ionic liquids, or molten salts. And it was already known that these molten salts would dissolve biomass, we knew that. The question was, can we turn that into a reasonable process. So, we made a pretty big investment in it, and I would say about a year-and-a-half ago – we, the management team, understand what's going on, we understand the science, we can see the progress, we know what else is going on in other institutions – we began getting pretty concerned that there was no fundamental breakthrough that was going to change the cost of using that technology. Our big concern was not, does it work, but could we ever afford it? Could it ever work, knowing that we have to meet the price of fuel? So, we started . . . pressuring the teams to say you know, they're doing all kinds of experiments of academic interest, but all this academic interest will go away if you can't find some route to get the price of the treatment to somewhere where we can afford it . . . fuels are so cheap that our processes have to meet a certain cost. And they're two orders of magnitude away from that cost point, and we're seeing the trajectory at which they're decreasing the cost and they're not really making progress on that core problem, though they are on all kinds of other stuff, right? (Quote from interview with EBI management team member, December 2011)

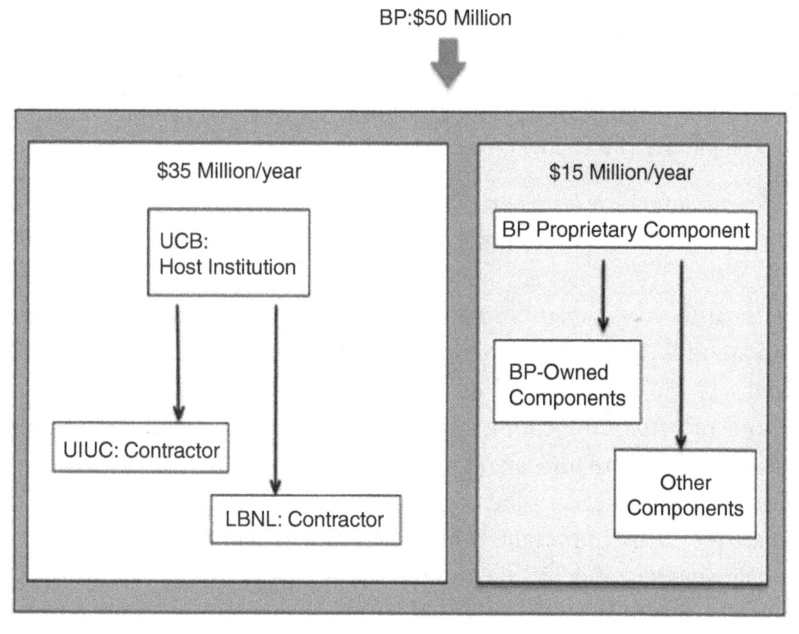

Figure 3.7. Funding and structure schematic for the Energy Biosciences Institute.

The management team decided to cut the program in the example, largely because the leaders at EBI saw the academic value of the research, but not usable technology as a result of market considerations. With their knowledge of the system, their willingness to integrate academic concerns into those of the market, and their knowledge about usability, the management sees its system as one that promotes strong academic research, but only when directed at addressing energy problems.

In addition to periodic meetings and frequent interaction with the management team, co-location can aid in keeping scientists, including those working on fundamental problems, aware of research output possibilities and eventual use. Both EBI locations, in UIUC and UCB, have active BP staff present in the labs, who interact with researchers on a daily basis. In addition, both UCB and UIUC possess lab spaces that researchers use as a shared resource, thus necessitating communication between research groups. The laboratory spaces do not include all EBI participants, but decision makers attempt to locate larger research programs in the shared labs, saying, "If we're looking at funding a major activity, we also make sure the work can happen in common space, for the purpose of getting people together and helping them to understand the whole chain." EBI sees it's a PI's participation as integral to meeting its mission. Thus, "we are not funding PI's and their laboratories, we are funding them to supervise people within the institute itself."

One interviewee, a plant biologist, described the benefits of such a system as more responsive research:

I used to be in a different building from the economists, I'd probably only meet with them once per month. Now I see them every day or every couple of days, at coffee breaks or wherever, and I can talk to them about what I'm working on, and sometimes they'll let me know if what I'm working on fits with the model they're working on, or if it doesn't make a difference to their model outputs. And then I can work more quickly to change that, if needed.

Although communication between laboratory groups does not ensure better results, it might allow for both new problem-solving approaches and more awareness of research context. In a *New York Times* account of Bell Labs, one author describes the physical architecture of the renowned laboratory, where the chairman of the board worked to "design a building . . . where everyone would interact with one another" (Gertner, 2012). Many of the interviewees, especially those with offices in the EBI-Berkeley building, did mention co-location explicitly as a benefit, including exchange with other researchers on both topics and methodologies, along with communication with BP staff, as valuable effects of the co-location.

EBI also plays an important role in training students. One estimate counted 138 graduate students and 157 post-docs funded through EBI. Regarding questions about interaction with BP and other EBI researchers, many students had had a high amount of interaction with cross-disciplinary EBI laboratory groups, with one directly citing the EBI laboratory setup at UCB for this. Interaction with BP was more varied. Interaction ranged, from zero or infrequent interaction at evaluation meetings, limited mostly to listening at seminars and preparing slides for presentation, to closer relationships. One student said that interactions with one BP employee ranged from "bouncing ideas off of him" to getting input on manuscripts to "see what's important from an industry perspective." One interviewee also described Sharon Rynders, a BP representative at UIUC, as "somewhat of a mentor to me." When asked about the differences between work at EBI and traditional academic departments, students said that the amount of collaboration within EBI, both within and across disciplines, seemed higher. They also cited a higher confidence in their funding sources.

3.4.3. EBI Performance

Given issues of funding and personnel management, the creation of EBI has helped UCB gain scientific expertise. From EBI, UCB not only received funding for existing researchers, but also for seven new hires in a necessary field, "the EBI provided resources for the startup money and allowed Berkeley to strengthen its program in scientific disciplines that are sometimes financially constrained in expanding or replacing faculty" (management team member). While UCB had expertise in biology and many other fields prior to EBI, it did not have any organized program in energy biosciences. EBI has allowed the university to build a program around this field of application, both encouraging UCB professors from related fields to work on applied problems in energy biosciences, and hiring new faculty to fill UCB's expertise gaps in the field. All three of the public institutions involved have benefitted from the funding supplied to them.

One concern related to a structure such as EBI is that it represents a move away from curiosity-driven university research and toward direct service of private profit, to the detriment of knowledge creation and public benefit. At EBI, the researchers are working on problems that cater to an industry's needs, but it is a research effort that pursues knowledge directed toward public interest problems, such as climate change mitigation, water and land footprint reduction, and diversification of fuel supplies. BP may harness some of these public values, with profit as a result, but this does not preempt them from also being a common good. BP may benefit disproportionally from the public institution's work, but the addition of these technologies to those available to consumers could be beneficial, if they can increase energy supply or reduce pollution. In addition, the fact that many of the outputs will not be proprietary and published as scientific articles, among other routes, but will lead to the advancement of strategic knowledge of the research area and non-proprietary intellectual property advancements, will result in more benefit for universities and the public.

BP gains two main benefits for funding the EBI. One of these is largely promotional because BP is providing funding to investigate a lower-carbon energy future. When BP announced its decision to fund EBI, its press release noted that it was a "significant step in BP's commitment to providing new lower carbon energy" (BP Press Release, 2006). The second is that of innovation, including knowledge that BP can use directly, in the form of patents. One interviewee estimated that EBI had already produced two or three patents "which are almost ready to be beneficial to them." In addition, the convening power of EBI, which has brought many more researchers into the field, all working under one mission, reportedly "means that BP has a better, more global view of the landscape for bioenergy."

The EBI will doubtless add new knowledge to the field for energy innovation. One such discovery lies in *Neurospora*, a fungus that EBI has studied that could hold important ramifications for learning how to degrade plant cell walls during biofuels processing. *Neurospora* could lead to a solution with the problem of the digestion of lignocellulose. The products of the digestion include two types of sugar, C5 and C6. The digestion process for C6 sugar has been known for a long time, and can be accomplished with yeast. However, researchers did not know how to digest C5 sugars until approximately 20 years ago, and that process is still a slow one. To do it on a large scale would involve the utilization of a large number of fermentors, with each running for a long period of time. One of the UCB–EBI researchers discovered that the organism *Neurospora* could digest C6 quickly. According to one interviewer, "Someone at BP saw a connection between the *Neurospora* work at Berkeley" and other work that was going on at UIUC, "and that led to a discovery on how to genetically alter the yeast to improve the fermentation rate." Interviewees cited the discovery both as a cross-disciplinary success and as one that integrated the expertise at

EBI's participant institutions. Such discoveries, which could streamline the processes of making biofuels production competitive with other fuel types, could be important steps in the creation of a biofuels industry.

Constraints, whether they are a result of congressional oversight of federal programs, regulatory barriers to an innovation, changing prices, competitive advantages from different technological outcomes, or administrative hurdles, can hamper the innovation process in any institution. EBI faces all of these constraints, but in some cases it may be able to circumvent some of them due to its structure. For example, federal laboratories often need to find corporations willing to fulfill the deployment aspect of innovation; BP often does not, and can partner or acquire businesses to do so when it is itself unable. An example of this was its 2009 joint venture with DuPont, Butamax, formed to commercialize biobutanol. In addition, administrative constraints are largely dependent on the organization itself. The EBI, with a fairly minimal management structure, and a leadership team that can pursue the research paths it desires with relatively short decision times, may benefit from its ability to operate with fewer decision nodes.

There are, of course, drawbacks to this system as well. The small management team also means that EBI must face the limitations in viewpoints that accompany it. One way to work around this, which it appears EBI has sought, is to bring in outside experts, through faculty recruitment, advising, and consultation with others, both at BP and outside. An additional drawback might lie in the fact that there is the potential for one market participant, BP, to benefit disproportionately from EBI discoveries. This is partly necessary, as BP is the only participant making large funding outlays to enable EBI, but BP may not see the benefit of a research path or technology that EBI pursues that a different entity might notice. EBI attempts to circumvent such risks through two mechanisms. First, while BP definitely informs the research process, EBI staff makes the decisions. Second, IP to which BP does not claim exclusive rights is available to UCB and the other participants in the institute. Such systems do not completely annul the potential drawbacks of the system, but do allow for a more even distribution of benefit and a more fully informed process.

EBI possesses many characteristics likely to aid it in pursuing useful energy innovation. First, its incorporation of systematic sociotechnical analyses of the problem, through its working groups and the incorporation of social, economic, and environmental research into its research program, promise to give EBI decision makers and researchers knowledge of context. Second, the close interaction it has with its main customer, BP, will likely enhance the ability of the two to build social capital and legitimacy. While successful science and technology is still dependent on credible and usable outputs, the processes that build trust between researchers and technology users are likely to improve the chance of uptake (Cash et al., 2002). Finally, EBI has the advantage of possessing an extremely stable budget, at least over a 10-year period, while at the same time being administratively lean, with leadership that has a level of independence in its decision-making. Such a situation might enable EBI to consistently follow promising research paths, but also to be flexible, following more promising routes, or cutting off research that might not lead to innovation. EBI's funding model is a special one, and may not be transferable to other innovation institutions. However, mechanisms that enable similar outputs, by integrating social and context analysis into the system, by building relationships between research performers and the user community, and by allowing for independence while providing stability, might benefit energy innovation institutions of all kinds.

3.5. Applying Lessons from SRC and EBI

Both EBI and SRC display a decision-making flexibility and accountability to customers. Although it might be difficult to incorporate this aspect fully in a federal institution (and although they are not to be emulated in their entirety for the purpose of reforming the national labs), there are some transferable lessons and strategies. Both institutions have been successful at building and maintaining relationships with a large portion of the industry that concerns them, and the direct relationship between evaluator, funder, and user at each can add effectiveness. For federal energy institutions, the funder will always be the government, but more responsiveness to its user groups – the firms that may commercialize and deploy technologies developed by the labs or in conjunction with the labs – may aid it in serving them.

Buy-in by the user group is ensured at both EBI and SRC because the institutions would not be able to exist without BP and the semiconductor industry, respectively. In contrast, the relationship between national laboratories and the named customer – industry – is not as tight, owing to the realities of DOE funding structures. The national laboratories could conceivably fail in their mission and continue receiving funding from appropriators if they are unaware of, or unconcerned with, such failures to adequately meet user needs. For this reason, building accountability to user groups is a necessity for the labs. Formal decision-making responsibility should always lie within the agency or the laboratory itself, as someone needs to be held accountable for the final decision, but institutions should allow inputs and guidance by including customers, in mechanisms that could be similar to those that SRC incorporates with its yearly reviews and those that are present on a daily basis at EBI, with BP employees located in-house. An advisory process that formally included members of the industries served by laboratory research could enhance the incentives for the laboratories to work toward industry benefit.

Because SRC membership is based on yearly dues, its funding is quite stable. The situation is such that funding will change drastically only if multiple members were either to leave or to join the organization. This stability is likely as long as SRC keeps its members satisfied. Likewise, EBI knows its funding is secure for at least the 10-year term. Federal institutions do not typically have this luxury.

SRC has successfully integrated academic and industry concerns for years, and in some of its programs, such as the FCRP and NRI, has included active participation of government entities as well. To an even greater extent, EBI marries the conduct of university research with both broad and the specific needs of a single industry along with the collaboration with a national laboratory.

In addition, SRC excels in its organizational flexibility. The administrative underpinnings of the institution are small, but it, along with the cooperation of its advisory boards and other member inputs, can change its research thrusts yearly to make new, more effective research policy when needed. EBI benefits from a small core of leadership that can make many decisions on its research programs while being insulated to an extant from interference; the governance board has little explicit authority, beyond the ability to hire and fire new members of the management team. Conversely, much of federal research direction is managed outside of the laboratory, in DOE, and in Congress. To allow them to adapt, NREL and other innovation institutions should have significant influence over their program direction. Placing more decision power with labs themselves, especially in the form of discretionary, lab-directed money, could aid them in responding to upcoming problems or opportunities within their fields, where they are both more expert and more able to respond quickly than the whole of DOE could.

EBI's lean management system does necessitate strong leadership but it also appears to benefit from its current leadership. At least, both BP and the partner institutions seem to be benefitting from the current arrangement, and the EBI governance board has made no attempt to modify Chris Somerville's leadership or to remove him from his position. Somerville's active role in steering the institute in a course that hews to EBI's greater mission, but is also responsive on a short-term basis to the input and direction of the management team, allows for an organization that is broadly consistent in its innovation approach, but flexible in how it operationalizes this approach.

The use of industry funding to produce a generation of young people expert in the semiconductor field is also integral to the success of SRC and EBI. SRC has put effort into building a base of graduate students to comprise its talent base. The number of U.S. dissertations with the descriptor "silicon" almost tripled, from 162 to 470, between 1982 and 1990. In the same time period, dissertations using the key word "integrated circuits" went from 31 to 110 (Burger, 2000). Many of these dissertations emerged from SRC funding. These students contribute valuable semiconductor research during their graduate careers, and many go on to work in the semiconductor industry, ensuring continued innovation. Similarly, EBI trains hundreds of students and post-docs, with at least some portion benefitting from interaction with industry employees.

The willingness of SRC and the semiconductor industry to come together and look at its future direction has proven to be one of its strengths. A 2010 President's Council of Advisors on Science and Technology (PCAST, 2010) presentation stressed the importance of technology roadmaps in its proposals; SRC's long participation in SIA road mapping is an important one. Examination of the road mapping process could lead to more insight for energy projects going forward with an eye toward having an impact on technology. The DOE roadmaps would have input from the external community and manifest "goals and milestones, critical decision points, projections of cost, benefits and analysis."

Finally, EBI's approach reflects one thing that is pursued at NREL, and should be at other institutions: outside of the pursuit of innovation as an end in itself, EBI uses some of its resources to pursue systemic understanding of the problems it addresses. Similarly, NREL uses SEA to address the socieconomic aspect of the energy problems it looks at. Such a willingness to devote effort not only to the creation of inventions, but also to developing a knowledge base surrounding the application of the inventions, can give EBI, NREL, and other institutions a knowledge similar to that developed by SRC in its road mapping. All work toward understanding where developments in the field may lead in the near and longer term future without minimizing the role of serendipity, face-to-face interactions, and other activities to encourage connections that cannot be foreseen in even the best roadmapping exercises.

3.6. Recommendations

NREL's problems relate to the need for the institution to maintain consistency and flexibility. Even given a stable budget, it often needs to alter its course to address changes in markets, policy regimes, and political realities. The lab's ability to address these challenges on its own is limited, owing to the authority of EERE, DOE, the executive branch administration, and the legislature. However, the laboratory can work within its boundaries to have an impact. NREL is already addressing this by working with user groups, encouraging CRADAs as well as other relationships with industry, and by attempting to address gaps between basic and applied research. However, budget variability, the level of administrative separation, the lack of formal connections between basic and applied research, and the lack of flexibility for researchers to conduct one- to two-year stays in other businesses and institutions, could all be remedied with different policy approaches. The recommendations below are suggested as a means for NREL, and energy innovation institutions in general, to improve their performance.

1. **Provide consistent institutional support.** A stable vision and guiding principles for laboratories can enable development of long-term projects, allow for planning beyond the next budget period, and enhance the likelihood that knowledge gains can be developed into eventual technology products. These principles should also be central to the new DOE innovation institutions, including ARPA-E, the Joint BioEnergy Institute, and the Energy Innovation Hubs. Three main activities can facilitate this.

a. **Encourage stable budgets.** Both SRC and EBI benefit from long-term budget stability. At NREL, the current, unstable situation both detracts from the ability of the laboratory to plan for the future and sends the message that the government does not consistently support the NREL mission. Without a more consistent approach, this trend will likely continue. For the federal organizations managing such laboratories, it is necessary to give the laboratories the stability of funding and mission that they need to perform consistently.

b. **Maintain a consistent mission and consistent implementation of mission.** NREL's mission, as worded, has not changed greatly over time, but external factors have altered the degree to which it prioritizes commercialization, advancement of knowledge, or both to fulfill its mission goals. Furthermore, the higher-level rhetoric surrounding renewable energy has changed and altered NREL's implementation of its mission. For example, George W. Bush stressed hydrogen and biofuels during his administration while President Obama has continued prioritizing biofuels, but not hydrogen to the same extent. In addition, while much of the laboratory receives its funding from EERE, and thus works under the umbrella of that mission, OS funding is not linked to the same goals. Lack of reconciliation between these higher level missions constrains the lab in ensuring that all of its research works toward its own mission.

c. **Manage contracting to allow for insulation of the national laboratory.** More complicated administrative operations are not a definite outcome of the contracting model that most national laboratories operate under, but they are a risk. Narayanamurti et al. (2009) maintain that the laboratories used to follow an "insulate but not isolate" principle, where the government funded and managed contracts, but labs were somewhat free to pursue their missions without frequent subjection to political considerations. They warn that changing political climates have altered contractor management of laboratories from an entrepreneurial system to one that resembles typical civil service institutions. One place where this could manifest is in contract management, where the terms of success and failure are negotiated. DOE management of such contracts to allow the laboratories significant independence and flexibility is necessary.

2. **Integrate the DOE energy research system.** In addition to the national laboratories, DOE now operates through a variety of mechanisms, including ARPA-E, Energy Frontier Research Centers (EFRCs), and Energy Innovation Hubs. Though support of overlapping, distinct thrusts in energy innovation can be useful, some amount of coordination between all of the actors in energy innovation is necessary. Though some of this coordination is taking place, more is needed. Coordination could include integrating the fundamental research projects performed at EFRCs with applied programs at Hubs and labs, inclusion of laboratory staff in Hub or EFRC activity, communication between ARPA-E and other institutions, and collaborative long-term planning. Such activities should include the three recommendations below:

a. **Support both fundamental research and technology development.** NREL is working to tie its basic research to applied research and development, but both EERE and OS need to ensure that such work does not go neglected. Part of the impetus behind basic research that eventually leads to good outcomes can be knowledge of and direction toward the field of application and recognition that the innovation ecosystem is bidirectional where feedback is an important ingredient for success (Narayanamurti et al., 2013). In addition to considering application-oriented fundamental research, programs must institute means for picking up where basic research leaves off, and supporting the project across the "valleys of death" that can ensue when such linkages go unsupported. OS funds much useful fundamental research, but should connect such research to an eventual application in an effort to enhance energy innovation. This should be an explicit part of its mission and operations. Though interaction and mission orientation are both important steps in this process, the inclusion of formal guidelines for such processes can benefit the laboratory. Examples exist in other federal institutions and nongovernmental research institutions. For example, DARPA, National Institute of Standards and Technology (NIST) (Logar, 2009), and private industries (Tennenhouse, 2004) have instituted the Heilmeier Questions, named after a former director of DARPA, that encourage consideration of technical demands along with the applicability of the project and fit to mission. One version of the Heilmeier Questions is:

What is the problem; why is it hard?
How is it solved today, and by whom?
What is the new technical idea; why can we succeed now?
Why should NREL do this?
What is the impact if successful and who would care?
How will you measure progress?
How much will it cost and how long will it take?

Usability need not be encouraged at the expense of long-term research. NREL must also manage to retain a balance that allows for support of some high-risk fundamental research. Given this risk, NREL management must be savvy enough to strike an appropriate balance between short-term applications and long-term, high-risk research.

b. **Continue focusing on strategic analysis of energy problems.** DOE currently supports such activities at several labs. It should also widely support planning for future innovation efforts through road mapping activities. Such activities have been proposed by members of PCAST, and are one place where lessons in research prioritization and management could be taken from SRC. Organized formal interactions between analysts, researchers, and decision makers can also increase fluency in energy contexts and opportunities.
c. **Continue supporting a portfolio of institutions supporting energy innovation.** The administration and Congress should provide support for a portfolio of energy innovation institutions designed to support all stages of the innovation chain, while avoiding gaps and "valleys of death." In particular, they should provide sustained support for ARPA-E, the Energy Innovation Hubs, and the Energy Frontier Research Centers and be patient with those new institutions, given the long timeframes involved in innovation. These institutions are described in more detail below.

3. **Develop a workforce for the future by funding mission-specific graduate and postdoctoral education.** Explicitly aim a portion of research dollars at encouraging research in energy innovation. SRC attributes much of its success to its ability to produce future researchers by funding students in its field. Some amount of DOE funding should be geared specifically toward training new professionals in energy science and technology, especially

in cases where cross-disciplinary or multidisciplinary experience is needed. Some programs, such as ARPA-E and the EFRCs, do fund university programs, but all units could benefit from more programs aimed explicitly at addressing graduate education within their program areas, and labs such as NREL could have strong results from mission-directed graduate research funding. Such programs for university research could serve as an important means for acquiring a long-term, sustainable, skilled workforce.

4. **Continue building relationships with user communities.** NREL published a report in 2003 that recommended "early and frequent interactions" between public sector institutions and the clean energy companies to promote use and commercialization of laboratory science. NREL has proven its technical competence through innovative research and the organizational benefits from a workforce that identifies strongly with its mission. There is ample evidence that NREL researchers and decision makers have a deep knowledge of their constituents, especially in areas such as thin film PVs, but more breadth in technology area and representation from companies in all areas of renewable energy could help the lab understand its innovation problems systematically and comprehensively. Exploring further developments, such as testing activities that bring industry groups to NREL to do work, could benefit the laboratory in much the same way co-location does at EBI.
5. **Encourage innovation through employee entrepreneurship.** The laboratories might encourage users to interact with their researchers, but allowing researchers to undertake entrepreneurial activities could also renew or complement their skills, improve the ability of the laboratory to learn from industry partners, and transfer technologies into the market. NREL allows employees to take leave, but only for six months, and these actions are not encouraged. Unlike NREL, Sandia has a program, called Entrepreneurial Separation to Technology Transfer, that allows researchers to be on leave from the laboratory for up to two years. According to one study on such programs, NREL's technology transfer staff argued "that because researchers are selected on their research skills rather than entrepreneurial talent or business acumen, they are not optimal founders of companies, as they neither have the interest nor the credibility from investors" (Kim, 2011). However, at Sandia, researchers do participate in the leave program, and 60% of them do so with preexisting companies, meaning that they may not need entrepreneurial or business skills. Rather, they may add necessary technical skills to existing entrepreneurial activities. Increased focus on allowing such activities at NREL could further aid it in transferring knowledge and expertise to business ventures.
6. **Expand methods for industry collaboration.** CRADAs are one way to promote partnerships and innovation, but DOE should explore other means of linking industry efforts to federal research, just as EBI has found ways to link its own efforts and interest to university research. CRADAs have had value; they have also been criticized for being slow and burdensome for industry partners. Though laboratory testing and Work for Others are useful, they do not enable the institution to effectively harness the expertise of its industry partners. Licensing is valuable when technologies are transferred, but occasionally a useful project can only be the result of partnership in concert on a research topic. Discussion of new ways to explore such relationships that can lower the hurdles for industry without allowing it to free ride on the laboratories' efforts could produce new means for partnering. In addition to bilateral partnerships, the labs should have the flexibility to contribute to multi-institution

centers devoted to addressing central problems of energy innovation. These centers could be virtual rather than physical, but should integrate strategic energy analysis and long-term thinking with research, technology development, and planning. Such centers might be similar to the Energy Innovation Hubs that have begun in the past few years, but should include the expertise of laboratory staff and their partners.

7. **Encourage strong relationships between researchers, people in commercialization, and industry.** EERE could support high-level interaction between NREL staff, DOE decision-makers, and technology consumers by increasing NREL's authority and accountability: NREL staff should be able to gain from social capital-enhancing activities such as procurement and contracting, as those that are close to the work being done have experience and input necessary for making the decisions. Managers should encourage strong relationships between researchers, people in commercialization, and industry.
8. **Increase funding for laboratory directed research and development.** This would provide the lab with a means for addressing needed research areas or pursuing high-risk research that EERE or OS is not funding. DOE Order 413.2B sets an upper limit for LDRD at 8% of a "laboratory's total operating and capital equipment budgets," but does not set a lower limit. Setting a minimum would ensure that a significant amount of money is available for pursuing unfunded but promising high risk research, proof of concept work, and development work on basic research that has paid off. According to decision makers at NREL, increasing the percentage at smaller labs is difficult because it must come out of direct funding and it dramatically drives up overhead costs. Therefore, augmenting LDRD amounts might require changes to the funding mechanisms. However, allowing for more base funding and manager-directed funding at the laboratory could create both consistency and flexibility at the lab, and allow for NREL leadership to take an active role in directing the course of the laboratory. NREL currently uses $8 million per year for laboratory-directed research and development, which is only 2.6% of its 2009 Congressional budget, and much lower than the 8% limit. A larger amount of discretionary funding could help the director fill holes in knowledge that are missed by higher level decision makers continue funding for promising areas that have been cut off, or explore new areas that are missed by current policy initiatives.

3.7. A Strategic Approach to DOE Energy Innovation Institutions

The principles just described, if implemented, should substantially improve the effectiveness of the national laboratories in accelerating energy technology innovation. The U.S. government should apply similar principles to all of its energy innovation institutions, taking a strategic, systems-oriented approach to maximize the return on its energy innovation investments. This applies in particular to the new elements that have been added in recent years.

The creation of these new institutions is a signal that U.S. policymakers realized that the previous model, relying to a large extent on DOE national laboratories and on an artificial separation between basic research funded by the Office of Science and applied research funded by the other DOE offices, was not producing the desired outcomes. As a result, the nation's energy innovation system is being examined, significantly expanded, and reshaped.

This reshaping brings not only a rare opportunity, but a responsibility, to improve the efficiency and effectiveness of this system. The importance of improving and better aligning the management and structure of existing and new energy innovation institutions to enhance the coordination, integration, and overall performance of the federal energy-technology innovation effort (from basic research to deployment) cannot be overemphasized. The technology-led transformation of the U.S. energy system that the administration is seeking is unlikely to succeed without a transformation of energy innovation institutions and of the way in which policymakers think about their design.

Various ingredients are needed for this transformation of energy innovation institutions. First, as described previously, the U.S. government needs to shift toward a more independent, stable, accountable version of the labs with significantly more interaction with private firms and universities. Second, a consolidation of ARPA-E and the Innovation Hubs would support high-risk projects and a sustained effort in areas of research requiring a critical mass of multidisciplinary researchers. Third, a demand pull policy (or policies) to support early commercialization in technologies that may be too risky for private firms but could materially impact the future of the U.S. energy system may provide such an environment. Below, we describe several of the new institutions and their potential roles.

Demonstration and Commercialization: Roles for Public–Private Partnerships[2]

The transition from the research and development phase to the commercial phase has long been recognized as a particular challenge for energy technology innovation. Demonstration projects – construction and operation of technologies at near-commercial scale (see more on demonstrations at the end of this box) – ease the transition by providing information to market participants. These projects are often too large or too risky to be done by the private sector, but too close to the commercial market to be conducted under normal government R&D support. In December of 2010, a group of senior representatives from government, industry, finance, and academia convened at the John F. Kennedy School of Government at Harvard University for a workshop to discuss

[2] The intent of a commercial scale demonstration is to show that a technology can be effectively used at full commercial scale in an economically useful way, that is, at a reasonable cost. However, such a scale-up is also the first opportunity to learn how a technology works at full scale, whether the design features are reasonably workable, controllable, and optimal. As such, the beginning of operation of a full-scale demonstration plant is an experiment, designed for learning and "debugging." A new plant needs adjustment, particularly in control systems and their use, sometimes alterations in hardware (hopefully minor), and always requires some accumulation of experience in operating the plant. It is, in fact, a late stage of development. It provides an opportunity to learn what changes might be made in the next plant(s) to be built. Such projects are not demonstrations of what is already fully understood, but experiments at the frontiers of knowledge and practice.

options for accelerating the commercialization of advanced energy technologies. A report on the workshop (Narayanamurti et al., 2011) describes the findings and policy principles that emerged from these discussions.

Current policies do not adequately address the private sector's inability to overcome the demonstration "valley of death" for new energy technologies. It has been argued that part of the reason for this is that some previous and prominent demonstration efforts at DOE have been widely perceived as failures, such as the Synthetic Fuels Corporation (Anadon & Nemet, 2013). Investors and financiers fear that the technological and operational risks at this stage of the cycle remain too high to justify the level of investment required to build a commercial-sized facility. Many participants agreed that demonstrations are particularly important to the energy innovation system because (1) they are currently a bottleneck and a rate-limiting step to the commercialization of energy technologies; (2) they test new business models that provide critical knowledge to different stakeholders; and (3) without a clear price signal, the need to support innovative energy technologies becomes more urgent.

Several policy principles and recommendations are outlined in the report. The most important of these are:

Long-term policy. The most consistent theme from the workshop was that if the government wants to unlock private-sector investment, it must create a predictable and sustainable policy environment.

Materiality. Government support for accelerating the commercialization of energy technologies should focus on projects that have the potential to have a material impact on one or more of the objectives of energy security, economic competitiveness, and environmental sustainability.

Public–private partnerships. Because the ultimate objective is to enable the private sector to move new capabilities into the market, government policy should emphasize working with industry, obtaining private-sector input, and enabling private sector investment.

Unproven technologies. Currently there are no permanent mechanisms for supporting the demonstration of unproven innovative technologies. Though some of these may "fail," even the failures provide useful lessons, and the potential benefit from those that do succeed would likely dwarf the results of a more risk-averse portfolio. Further, it is the unproven as opposed to proven technologies that benefit most from government support.

Information dissemination. There is a clear public interest in making the information acquired from demonstration projects more widely available, although this will need to be balanced with the protection of private intellectual property rights to ensure private-sector participation in these projects.

Exit strategy. Government support for technology demonstration and deployment projects should be conditional on performance and based on transparent performance benchmarks, and if these are not met, funding should be discontinued. In other words, the government should have a clear strategy for ending its support of under-performing projects.

(*continued*)

Targets and objectives. Targets for demonstration programs should be based on technical performance criteria as opposed to far-reaching, large-scale production targets, which can be expensive and inefficient because their achievement often depends on technological advances and commodity prices.

Portfolio approach. The government should adopt a portfolio approach that supports a range of competing technologies, a range of development stages, and a variety of policy and financial mechanisms. However, given the large scale of investments, not all projects and technologies can be supported, and real-world constraints dictate a need to be selective. For instance, some workshop participants prioritized support for CCS and nuclear energy due to their potential to address overarching policy objectives, and their high cost, which makes it very difficult to attract private financing.

There was widespread, though not unanimous, agreement that a new institutional mechanism was needed to manage demonstration projects in partnership with the private sector, that existing institutional approaches at DOE were not likely to be effective for that purpose. In particular, there was broad, although not universal, support for two policy initiatives currently being discussed by policymakers. One is the Quadrennial Energy Review (QER), which was proposed in 2010 by the President's Council of Advisors on Science and Technology (PCAST) to bring much-needed coherence, stability, and predictability to energy policy in the United States (PCAST, 2010). The other is the Clean Energy Deployment Administration (CEDA), which was viewed by many as a useful policy tool for enabling energy technology demonstrations. The fact that CEDA has already received bipartisan support in the Senate was an important factor in the support for the workshop's own initiative.

3.7.1. ARPA-E

ARPA-E was established by the America COMPETES Act of 2007 after a recommendation from the National Academies' influential *Rising Above the Gathering Storm* report (NAS, 2005). ARPA-E was created to fund high-risk, high-payoff energy RD&D that both the DOE system and the private sector might otherwise miss, and was first funded in 2009 through the Recovery Act.[3] ARPA-E received $388 million for its first two years of operation and then $166 million in fiscal year (FY) 2011 and $255 million in FY 2012. The substantial budget increase for FY 2012 is a testament to ARPA-E's hard work and success in convincing legislators of its potential. Nevertheless, some legislators have been calling for its budget to be cut since its inception,

[3] Modeled after DARPA, a research agency at the United States Department of Defense established in 1958 to maintain technological superiority in the United States military. Its focus is to take technologies to the marketplace by funding (rather than conducting) research and aims to (1) bring "freshness and new sense of purpose" to energy research; (2) provide a mechanism for funding high-risk, high payoff (or transformational) energy research that has not been adequately supported by the private sector or DOE efforts; and (3) overcome the bureaucratic limitations of DOE, which was perceived as being unable to attract the brightest minds, sustain promising projects, and terminate unpromising programs (Chu, 2006).

and continue to do so; this highlights the difficulty of convincing U.S. policymakers to allow time for transformational innovation goals to be accomplished.

ARPA-E provides funding for research projects in the private sector, universities, and national labs at a scale of $0.5 to $10 million per year over two to three years, along with support to help gather follow-on funding and test technologies in market environments. Like DARPA, its investments are shaped by ARPA-E program directors, who are selected for their technical qualifications (their understanding of the underlying scientific and engineering problems to be addressed, which are decided by the ARPA-E director) as well as their knowledge about private sector demands.

As highlighted by Bonvillian and Van Atta (2011), DARPA was successful in part because in many instances it worked over extended periods of time to meet particular goals; some of its projects brought together families of emergent technologies in ways that transformed the technology landscape (examples including the internet, global positioning satellites, stealth technology, and the technologies that enabled stand-off precision strikes). In the process of doing this, it created technical communities that could take technologies forward. ARPA-E is trying to replicate this. Its success in doing so will depend on its ability to manage the much more complex environment of developing cost-effective technologies for private markets (rather than having a single Department of Defense customer for whom the mission was more important than the initial cost, as DARPA did), and on its ability to maintain support over time.

ARPA-E, like DARPA, has some of the important features that the labs often lack, such as greater insulation from DOE (even though it is part of the DOE structure, the ARPA-E Director reports directly to the Energy Secretary), the ability to quickly hire and attract excellent program managers, and the ability to release funding quickly for high-risk, high-payoff projects, but it still does not cover the full spectrum of research needed to meet U.S. energy challenges – which is why DOE has created other institutions as well.

3.7.2. The Energy Innovation Hubs

Like ARPA-E, the idea for the Energy Innovation Hubs also mainly originated from the academic community, with the realization that the research that was taking place was not tackling some key energy challenges with a sufficient critical mass of interdisciplinary researchers (Anadon, 2012). Secretary of Energy Steven Chu, who had been at the AT&T Bell Laboratories before becoming a professor and national lab director, was responsible for the creation of the Energy Innovation Hubs. He was inspired by the Manhattan Project, MIT's Lincoln Labs, AT&T's Bell Laboratories, and the DOE Bioenergy Research Centers described above, all of which he saw as being intensely mission-focused and seamlessly integrating basic and applied research. He was also influenced by a Brookings Institution proposal led by Prof. James Duderstadt,

President Emeritus of the University of Michigan (Duderstadt et al., 2009). The Hubs bring together a large set of multidisciplinary investigators under one roof to focus on one general topic area, and have sustained funding of about $25 million per year over 5 years. There are Hubs on nuclear modeling, fuels from sunlight, batteries for vehicles, more efficient buildings, and critical materials. It is notable that all five Hubs have a national lab as a partner, and in three of the Hubs a national lab is the leading organization. This will be positive if the Hubs work as designed – with a laser-like focus on their mission, with technically excellent multidisciplinary teams, more stable and long-term funding, increased independence and accountability – and if over time they shape the management and work that gets conducted in the participating labs and other labs.

3.7.3. The Energy Frontier Research Centers

The EFRCs have their roots in DOE's Basic Energy Sciences Advisory Committee (BESAC), which is mainly composed of academics and National Lab members. A series of workshops organized by BESAC between 2002 and 2007 coalesced around five major science and engineering challenges, which were outlined in a 2007 report (BESAC, 2007). This process led to the concept of the EFRC, a multi-investigator center working on one area of research falling under one of the five challenges, selected by peer-review, and funded at $2 to $5 million for a 5-year period. In general, the work of the EFRC's is further from commercial application than the work supported by ARPA-E and the Energy Innovation Hubs. There were 46 EFRCs established in 2009, two-thirds of them hosted at universities. As with the Hubs and ARPA-E, the first funding for the EFRCs came from the Recovery Act, amounting to $777 million over 5 years. In essence, the stimulus package provided an opportunity for DOE to finance new programs without cutting back in other areas – but that approach has created challenges for future funding.

As with ARPA-E and the Hubs, the EFRCs main proponents and implementers came mostly from the academic and national labs communities (Anadon, 2012). In addition to increasing the support for strategic areas of science and engineering, the EFRCs are training a cadre of younger researchers (as of the time of writing around 2,000 students, postdoctoral fellows, and staff are involved in the EFRCs).

3.7.4. Still Missing: Institutions for Technology Demonstrations

As the box on p. 116 in this chapter suggests, for some technologies there is a compelling need for some government support for large-scale technology demonstrations – when the technology is at a stage where it is almost ready for commercial deployment but the risk is still too high for the project to be financed in private markets (Narayanamurti et al., 2011). Such demonstrations are likely to be expensive – potentially in the range of billions of dollars for a single project – and the government and the private sector are both likely to have to contribute. Though DOE has pursued a number of such demonstrations in the past, it has not yet done so successfully, and none of DOE's innovation institutions are well suited to the close partnership with the private sector likely to be needed.

Over the years, a number of concepts for such an institution have been proposed (reviewed in Narayanamurti et al., 2011). Congress has considered bipartisan legislation that would have established a Clean Energy Deployment Administration (CEDA), sometimes known as a "green bank," which would have helped to finance and manage such projects, as well as initial deployments of new technologies, until they became commercially competitive. At this writing, however, no such institution exists – and technologies such as new nuclear reactor designs or large-scale systems for carbon capture and sequestration are likely to suffer as a result of this gap in the U.S. energy innovation system.

3.8. Conclusions

The recommendations in this chapter can help the U.S. government build a more effective system of energy innovation institutions. Of the CASCADES criteria for an effective energy innovation strategy outlined in Chapter 1, this chapter emphasizes in particular the need for these institutions to be *strategic*, with a clear mission focus; managed in a way that is *sustainable*, with consistent funding over time; and *agile*, giving management of these institutions the flexibility needed to seize new opportunities as they arise, and having the risk tolerance needed to invest in projects that may fail and cut them off if they do.

NREL, EBI, and SRC are only three institutions, but from such cases, along with data on other types of institutions and experience from other labs (Logar & Conant, 2007; Logar, 2009; Narayanamurti, Anadon, & Sagar, 2009) we can begin to build information on how to manage innovation. Given the size of the energy economy and the pressures to reform our energy system, the need to continue innovating is high. While pumping more money into the U.S. innovation system would likely accomplish more innovation, programs that deliver results through effective management, long-term vision, and consistency of purpose could do much to improve the state of our innovation system. While such work only creates options, and these options can be made moot by regulation, market pricing, behavioral issues, or other unpredictable futures, working to deliver these options and alternatives should be a goal of any research institution.

References

Anadon, L.D. (2012). "Missions-oriented RD&D institutions in energy between 2000 and 2010: A comparative analysis of China, the United Kingdom, and the United States." *Research Policy*, 41(10):1742–56.

Anadon, L.D., & Nemet, G.F. (2013). "The U.S. Synthetic Fuels Corporation: Policy consistency, flexibility, and the long-term consequences of perceived failures." In *Energy Technology Innovation: Learning from Successes and from Failures*, pp. 257–74, edited by A. Gruebler and C. Wilson. Cambridge: Cambridge University Press.

Apte, P., & Scalise, G. (2009). "The recession's silver lining." *IEEE Spectrum*, 46(10):46–51.

Bonvillian, W., & van Atta, R. (2011). "ARPA-E and DARPA: Applying the DARPA model to energy innovation." *The Journal of Technology Transfer*, 36(5):469–513.

Bozeman, B. (2000). "Technology transfer and public policy: A review of research and theory." *Research Policy*, 29:627–55.

BP Press Release. (2006, June 14). "BP pledges $500 million for energy biosciences institute and plans new business to exploit research." London: BP.

Brown, G. (1992). "Guest comment: The objectivity crisis." *American Journal of Physics*, 60(9):779–81.

Burger, R.M. (2000). *Cooperative Research: The New Paradigm*. Semiconductor Research Corporation. Unpublished manuscript.

Cash, D.W, Clark, W., Alcock, F., Dickson, N., Eckley, N., & Jager, J. (2002), *Salience, Credibility, Legitimacy and Boundaries: Linking Research, Assessment, and Decision Making*. Cambridge, Mass.: John F. Kennedy School of Government, Harvard University.

Chu, S. (2009). Testimony of Secretary of Energy Steven Chu, *FY 2010 appropriations hearing*: 1–12.

DOE. (2008). "DOE selects ASE to manage and operate its National Renewable Energy Laboratory." Retrieved from http://energy.gov/articles/doe-selects-ase-manage-and-operate-its-national-renewable-energy-laboratory (Accessed February 25, 2014).

Duderstadt, J., Was, G., McGrath, R., Muro, M., Corradini, M., Katehi, L., Shangraw, R., & Sarzynkski, A. (2009). *Energy Discovery-Innovation Institutes: A Step Toward America's Energy Sustainability*. Washington, D.C.: The Brookings Institution; 78 pp. retrieved from http://www.brookings-tsinghua.cn/~/media/Files/rc/reports/2009/0209_energy_innovation_muro/0209_energy_innovation_muro_full.pdf (Accessed July 25, 2013).

EBI. (2007). Master Agreement, Dated November 9, 2007, Between BP Technology Ventures, Inc. and the Regents of the University of California, Energy Biosciences Institute, Berkeley.

EERE. (2006a). Performance evaluation of the Midwest Research Institute for the management and operation of the national renewable energy laboratory; October 1, 2005 through September 30, 2006 performance period. 2006:1–34.

EERE. (2006b). "President Bush promotes the advanced energy initiative." *Solar Energy Technology Program News*, February 22, 2006.

EERE. (2011). "Entrepreneur in residence program." Retrieved from http://www1.eere.energy.gov/commercialization/entrepreneur_in_residence.html. (Accessed July 17, 2011).

First Solar. (2003). First Solar's breakthrough technology wins R&D 100 award. *First Solar Press Release*. September 9, 2003.

Funtowicz, S.O., & Ravetz, J.R. (1993). "Science for the post-normal age." *Futures*, 25(7):735–55.

Gertner, J. (2012). *The Idea Factory: Bell Labs and the Great Age of American Innovation*. New York: Penguin Press.

Gibbons, M., Limoges, C., Nowotny, H., Schwartzman, S., Scott, P., & Trow, M. (1994). *The New Production of Knowledge: The Dynamics of Science and Research in Contemporary Societies*. London: SAGE.

Guinnessy, P. (2009). "U.S. weapons labs determined to retain funding." *Physics Today*. Retrieved from blogs.physicstoday.org/politics/2009/08/us-weapons-labs-determined-to.html (Accessed May 5, 2011).

Kim, H. (2011). *Commercializing Low-Carbon Energy Technologies from Research Laboratories through Start-Ups*. Unpublished MS dissertation, Oxford University.

Laird, F.N. (2001). *Solar Energy, Technology Policy, and Institutional Values*. Cambridge: Cambridge University Press.

Logar, N. (2009). "Towards a culture of application: Science and decision making at the national institute of standards & technology." *Minerva*, 47:345–66.

Logar, N., & R. Conant, R. (2007). "Reconciling the supply and demand for carbon cycle science in the U.S. agricultural sector." *Environmental Science & Policy*, 10(1):75–84.

Marburger, J.H., III. (2005). "Wanted: Better benchmarks." *Science*, 308(5725):1087.

Marczewski, R. (1995). Testimony of Richard Marczewski, manager, technology transfer office, National Renewable Energy Laboratory, for the House of Representatives Science, Space, and Technology Committee's Subcommittee on Technology and Subcommittee on Basic Research. House Report 104-390 – National Technology Transfer and Advancement Act of 1995, pp. 151–62.

Moniz, E., & Savitz, M. (2010). *Accelerating the Pace of Change in Energy Technologies through an Integrated Federal Energy Policy*. PCAST report, September 2, 2010. Washington, D.C.: President's Council of Advisors on Science and Technology.

Narayanamurti, V., Anadon, L.D., Breetz, H., Bunn, M., Lee, H., & Mielke, E. (2011, February). *Transforming the Energy Economy: Options for Accelerating the Commercialization of Advanced Energy Technologies*. Cambridge, Mass.: Report for Energy Technology Innovation Policy Research Group, Belfer Center for Science and International Affairs, John F. Kennedy School of Government, Harvard University.

Narayanamurti V., Anadon L.D., & Sagar A.D. (2009). "Transforming energy innovation." *Issues in Science and Technology*, 26(1):57–64.

Narayanamurti, V., Odumosu, T., & Vinsel, L. (2013). "Rip: The basic-applied research dichotomy." *Issues in Science and Technology*, XXIX.2, Winter.

NREL. (1995). "NREL funding reductions to further impact lab's workforce." *NREL News Releases*.

NREL. (2009). "Recovery act money invested at NREL." *NREL Feature News*, May 1, 2009.

NREL. (2010). "NREL's clean energy forum attracts nearly 500 participants from investment community." *NREL News Releases*, October 21, 2010.

OSTP. (1983). *Report of the White House Science Council Federal Laboratory Review Panel*, pp. 1–36. Washington, D.C.

Packard, D. (1983). "Packard report urges more autonomy without change." *Nature*, 304:199.

PCAST. (2010). *Report to the President on Accelerating the Pace of Change in Energy Technologies through an Integrated Federal Energy Policy*. Washington, D.C.: President's Council of Advisors on Science & Technology.

Pielke, R.A., Jr. (1995). "Usable information for policy: An appraisal of the U.S. global change research program." *Policy Sciences*, 38:39–77.

Pielke, R.A., Jr., & Byerly, R., Jr. (1998). "Beyond basic and applied." *Physics Today*, 51(2):42–6.

Pielke, R.A., Jr., & Sarewitz, D. (2003). "Wanted: Scientific leadership on climate change." *Issues in Science and Technology*, Winter: 27–30.

Rubin, H. J., & Rubin, I.S. (1995). *Qualitative Interviewing: The Art of Hearing Data*. London: SAGE.

Sarewitz, D., & Pielke Jr., R.A. (2007). "The neglected heart of science policy: Reconciling supply of and demand for science." *Environmental Science & Policy*, 10(1):5–16.

Serface, J. (2009). "My year as NREL's entrepreneur in residence." *Cleantech Blog*.

Shapley, D., & Roy. R. (1985). *Lost at the Frontier: U.S. Science and Technology Policy Adrift*. Philadelphia: ISI Press.

SRC. (2011a). "SRC vision, mission, charter and values." Retrieved from http://www.src.org/about/mission/ (Accessed July 17, 2011).

SRC. (2011b)."ERI mission and objectives." Retrieved from http://www.src.org/program/eri/about/mission/ (Accessed July 17, 2011).

Stokes, D.E. (1997). *Pasteur's Quadrant: Basic Science and Technological Innovation*. Washington, D.C.: Brookings Institution Press.

Tennenhouse, D. (2004). "Intel's open collaborative model of industry – university research." *Research-Technology Management*, (July-August):19–25.

U.S. Patent and Trademark Office. (2005). "The national medal of technology and innovation recipients." Retrieved from http://www.uspto.gov/about/nmti/recipients/2005.jsp (Accessed July 24, 2011).

Wiser, R., Bolinger, M., & Barbose, G. (2007). "Using the federal production tax credit to build a durable market for wind power in the United States." *The Electricity Journal*, 20:77–88.

Vance, A., & Miller, C.C. (2010). "6 months, $90,000 and (maybe) a great idea." *The New York Times*, February 27, 2010.

4

Encouraging Private Sector Energy Technology Innovation and Public–Private Cooperation

CHARLES JONES, LAURA DIAZ ANADON, AND
VENKATESH NARAYANAMURTI

4.1. Introduction

Private sector firms are critical players in energy technology innovation. As described below, the private sector funds and carries out the majority of energy research, development, and demonstration (RD&D) in the United States. New energy technologies can make significant contributions to major energy challenges only if they are adopted on a broad scale by the private sector.

In this chapter, therefore, we explore two sets of questions. First, what kinds of energy innovation are taking place in the private sector in the United States today, and how can the federal government best encourage them to do more? Second, how can U.S. government energy RD&D partnerships with the private sector – a major part of the federal investment in energy RD&D – be made more effective? As we show in this chapter, improving the U.S. Department of Energy (DOE)'s work with the private sector is important not only because the private sector will ultimately take technologies into the market, but because collaborative work between DOE and others makes up about 30% of the energy RD&D budget and 55% of the Science budget through grants and collaborative agreements. Within energy RD&D especially, for-profit firms receive 60% of all funding for collaborative work, and firms contribute more than they receive to jointly funded projects, so that firms represent 65% of the total collaborative effort.

As one of the world's wealthiest nations (measured in gross income per capita), the United States is well positioned to invest in the kind of R&D activities that drive innovation and foster economic growth. Spending on R&D by industrialized countries correlates positively with income, as illustrated in Figure 4.1; as wealth increases, nations typically spend more on R&D. But Figure 4.1 also shows that some countries with strong "innovation economies," such as Japan, Israel, Sweden, and Finland, spend more intensively on R&D than the United States even though they have lower per capita incomes. These data suggest that an opportunity exists to increase the intensity of U.S. R&D spending.

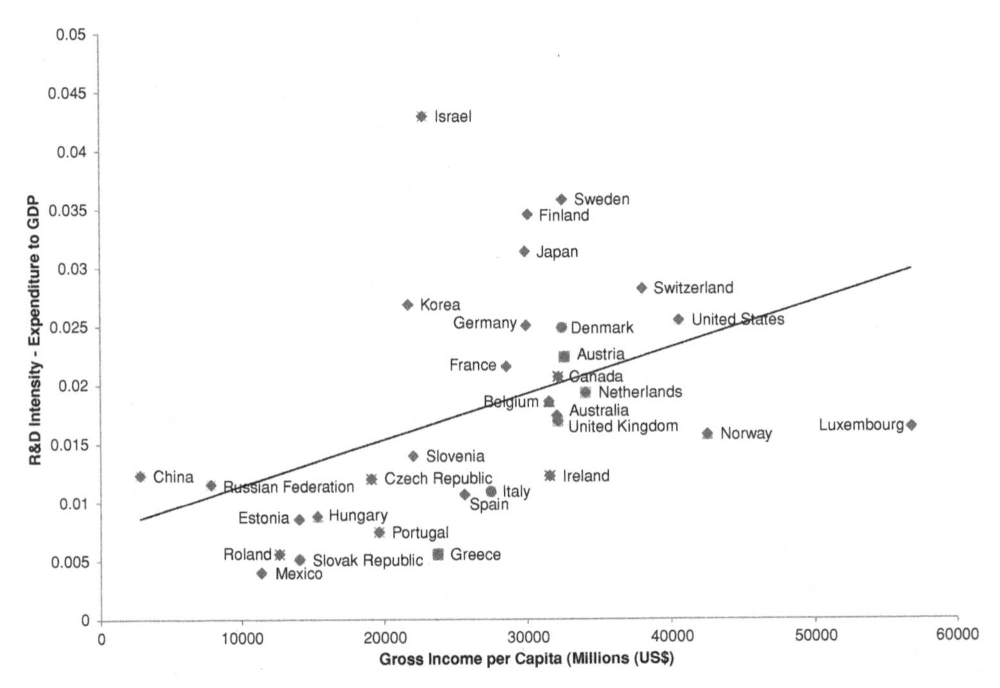

Figure 4.1. R&D intensity (R&D investment/GDP) and gross income per capita for a range of countries in 2009. (Data source: OECD Research and Development Statistics.)

Since 1990 the intensity of U.S. R&D spending has been roughly constant, measuring between about 2.5% and 3% of GDP (Figure 4.2). Most other large industrialized countries have also maintained a fairly constant level of R&D intensity. However, in the past decade R&D intensity levels in Japan and Germany have surpassed those in the United States. China's R&D intensity, although significantly lower, has approximately doubled since 1996. Important U.S. competitors clearly view R&D spending as a driver of economic growth.

Government, academic, and private sector research all have roles to play in energy technology innovation. In general, work by academic researchers and government laboratories dominates the early stages of discovery. The private sector plays an increasingly important role as development leads toward commercialization (Gallagher, Holdren, & Sagar, 2006). Both private firms and governments often sponsor research in universities, and governments often support innovative activities in private firms and otherwise influence private companies' decisions about innovation. Encouraging more private sector effort has been a goal of public policy in energy and innovation in general (DOE, 2006, 2011; CCTP, 2006; EOP, 2009). Though DOE is the largest single investor in energy technology innovation, the private sector collectively funds and carries out the majority of energy RD&D in the United States. However, as illustrated

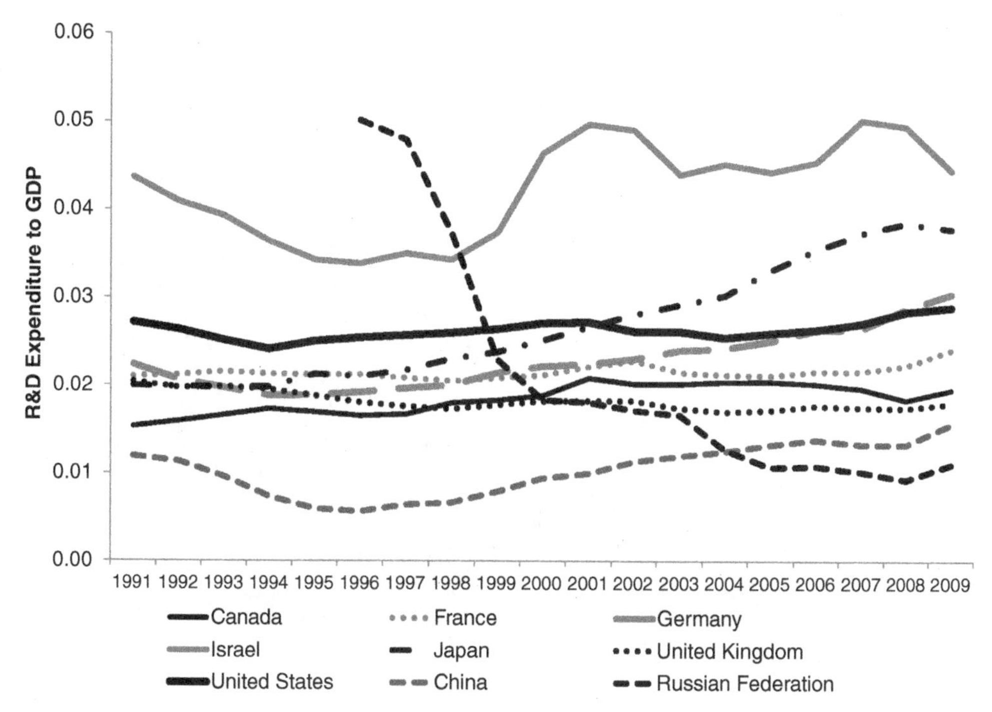

Figure 4.2. R&D intensity from 1991 to 2009. (*Source*: OECD Research and Development Statistics.)

in Figure 4.3, the public sector's share increased during economic downturns in 2001 and 2008, rising to roughly one-third of total R&D expenditures as of 2009. In 2009, a total of $400 billion worth of R&D was performed across all sectors in the United States.[1] Of the $400 billion, private-sector firms *performed* $282 billion (71%) and *funded* $225 billion (62%), or about 1.9% of gross domestic product (GDP) (NSF, 2012). The private sector overwhelmingly concentrates on the later stages of innovation: 6% of its funding went to basic research, 14% to applied research, and 80% to development. The result is that 90% of all development in the United States is funded by the private sector (NSF, 2013).

Investments in energy technology innovation are more effective when organizations collaborate and communicate across sector boundaries. Private sector innovation benefits from services, facilities, and information provided by the national laboratories. Government research benefits from the skills and resources of private-sector partners. Funding can be better allocated if funders understand the potential of technologies and the barriers being faced by the companies creating them. Many important projects require skills and resources that no one organization can provide, making partnerships

[1] This value represented a 1.7% decline from 2008, caused mainly by the financial crises.

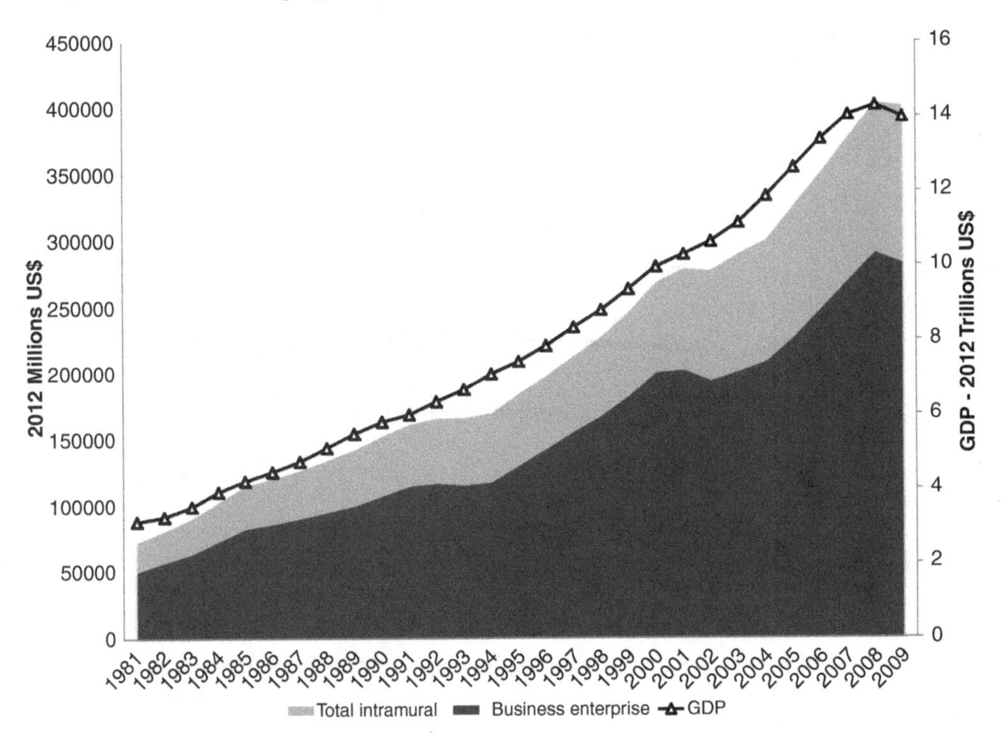

Figure 4.3. U.S. business and intramural (government) R&D (and GDP (right-hand-side axis) between 1981 and 2009.

the best strategy. Collaboration and communication across sector boundaries is vital.

According to current estimates, total private investments in energy R&D represent less than 1.3% of expenditures on energy, which is a much lower share than the ratio of R&D to total sales in other industries. (We address problematic aspects of these figures in Section 4.2.2.) DOE funding represents 39% of these rough estimates of total public and private funding in the United States. Grants and cooperative agreements with private firms represent about 30% of the energy RD&D budget and 55% of the Science budget within DOE. In 2011 DOE concluded its first Quadrennial Technology Review (DOE, 2011), an effort to formulate a new strategy to set priorities to invest available DOE RD&D funds effectively. It emphasized that mere technical promise was insufficient to justify the commitment of limited RD&D funds and that it is vital to understand more about the role of public–private partnerships and the private sector in energy technology research and development.

In this chapter we first offer data from a new *Survey of Energy Innovation* conducted for this study, which assessed both the scope of private sector energy innovation and key drivers of these investments. Based on the insights from that survey, we offer recommendations for U.S. government policies to strengthen private incentives to invest in energy technology innovation. We found that energy technology innovation is widespread, particularly among smaller-scale participants who had not been measured previously. Our survey was a pilot-scale study and should be expanded to provide better estimates and more detail. However, our results underscore the need for more strategic support for private sector innovation by the U.S. government. Increasing these investments, and using demand-side policies to increase incentives for private sector innovation through demand-side policies, would produce enormous benefits.

With an estimated 10.8 million active establishments in our sample pool, our results indicate that at least 120,000 businesses are currently involved in energy technology innovation (each of which is a potential partner for government energy RD&D programs), and that at least 250,000 businesses could benefit from developing or the development of new energy technologies. Demand-side policies, such as adopting a clean energy standard or setting a price on carbon emissions, would certainly increase the number of entities engaged in energy innovation and expand the efforts of firms that are already involved. Innovation is particularly important for startup energy companies, which report that they invest 75% of their startup capital in innovation-related activities (mainly in RD&D projects). Cost-related issues, including energy prices, are the most important drivers of innovation. The most significant barriers are the lack of markets for some energy technologies at current costs and the fact that the energy technology innovation system involves a multiplicity of actors (public, private, individual consumers, financial institutions, etc.) and requires long times and large investments. This chapter also analyzes DOE efforts that involve funding energy technology innovation by the private sector and universities. From 2000 through 2009, DOE awarded $9.1 billion to support basic science and $7.2 billion to support applied energy RD&D. Energy RD&D projects attracted $9.5 billion in outside matching funds, which demonstrates that private firms value these projects. The distribution of public funding for RD&D in the private sector is perceived as being volatile over time, and information about the outcomes of RD&D projects in the private sector funded by the government is limited, making it difficult to learn over time about how to best manage limited resources. We use public data, management lessons, our own survey of private sector energy innovation in the United States, and expert interviews to uncover opportunities for improving the effectiveness of public and private-sector investments through project management, strategic planning, and learning and adaptation.

We conclude that new strategies would promote more productive relationships between DOE and the private sector. Options include promoting more partnerships in strategic decision making; creating opportunities for learning and adaptation to improve project management; and improving DOE's relationships with and understanding of industry. These changes will help get the most value from investments in energy technology innovation and speed the transformation of the nation's vital energy systems.

4.2. Background

To be useful in market economies, products must be produced and sold by the private sector. Research conducted by universities or national laboratories risks being wasted if it does not transfer to firms that are able to commercialize discoveries (FLC, 2007a), or does not at least inform subsequent research efforts. Large private research labs once dominated the technology development process from discovery and invention (Narayanamurti et al., 2013) to commercialization, with basic research at universities acting as a reservoir of common knowledge. Increased deregulation in energy markets, pressures from shareholders, and globalization have changed the innovation landscape in the private sector, making it harder to pursue long-term research needed to meet the energy challenges outlined in Chapter 1. A government policy of merely supporting basic research and waiting for commercial results is no longer effective (Block & Keller, 2008; Mowery, Nelson, & Martin, 2010). But identifying what the government could do to fill the gap and encourage much needed technology innovation in energy in the private sector is compounded by two important difficulties: first, the amount and distribution of private investment is uncertain (as we discuss in Section 4.2.2, the evidence available suggests that investment levels fall short of what is needed); and second, as we point out throughout this chapter, the information available about the outcomes of previous projects to guide government decisions about partnering and funding private actors is insufficient.

4.2.1. Uncertainty in Energy Technology Innovation Investment

Information on private-sector R&D investments that specifically targeted energy technologies is very limited, and the data that were available until recently come from surveys that do not consistently cover the same sets of companies or industrial subsectors. Figure 4.4 shows the data collected by the National Science Foundation (NSF) Survey of Industrial Research and Development (SIRD), which asks a subset of large R&D performers (between 45 and 144 companies between 2000 and 2007) how much they invest in energy. The last time this survey was administered, it found that the 144 firms that were covered performed $5.7 billion of energy R&D (of which $5.3 billion were funded by companies).[2] It also includes data from the successor of SIRD, the NSF *Business R&D and Innovation Survey* (BRDIS). The BRDIS asked a sample of approximately 43,000 companies (which represented a population of approximately

[2] There are two other sources of data on U.S. private investment in energy RD&D. The EIA collected data on R&D expenditures for the largest major 27 energy companies, predominantly oil and gas for 2008 (EIA, 2009). These companies reported investing $2.2 billion in energy, of which $1.3 billion were invested in oil and gas recovery research. Dooley (2010) added data from EPRI and the Gas Research Institute to the NSF SIRD and reported $3.4 billion for energy RD&D for 2005 (this again, represents a low value). Like the NSF data, these studies represent a low bound, given the limited number of companies included.

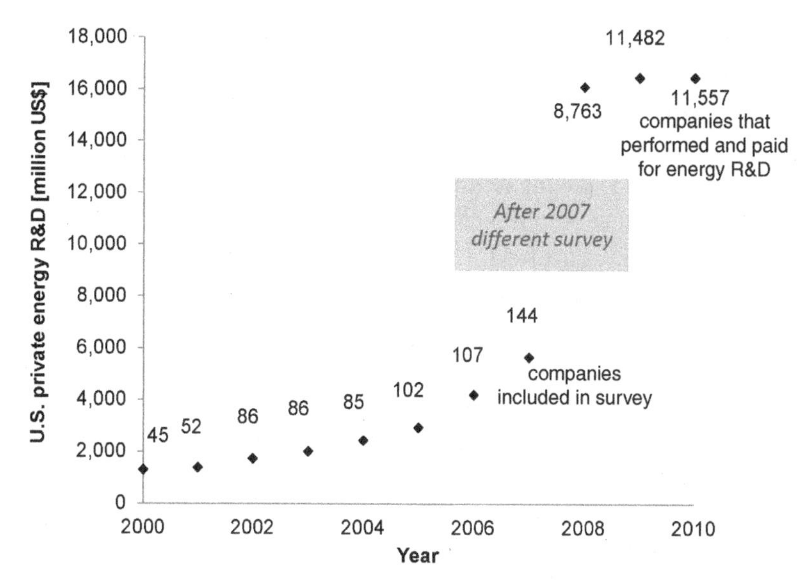

Figure 4.4. Energy R&D in industry from the NSF SIRD survey (from 2000 to 2007) and the BRDIS (for 2008–2010). The number of companies included in each of the surveys is noted next to each data point between 2000 and 2007. After 2007 the number of companies refers to the number of companies that performed and paid for energy R&D. (Data source: NSF, 2013, National Center for Science and Engineering Statistics, Business Research and Development and Innovation: 2008-2010, NSF 13-332.)

2 million companies in the United States in 2010) two questions about what fraction of their R&D effort has energy applications.[3] The NSF BRDIS showed that at least 11,500 firms performed and paid for R&D with energy applications in 2010, and about 2,900 firms that performed energy-related R&D paid for by others (there is some unknown overlap between these two categories). This means that about a quarter of the 48,000 U.S. firms active in R&D are working on energy-related R&D. Although the BRDIS highlighted the extent to which energy is important for firms, it did not provide insights regarding the technology focus of those investments and the policies that would be most effective at stimulating greater activity. In addition, the definition of energy applications would benefit from more specificity, which is admittedly difficult to achieve in a survey that is not energy specific and instead aims to cover all industrial R&D.

[3] The exact questions in the 2010 BRDIS read:

"Q4.12. What percentage of the amount reported in Questions 4–9 [domestic R&D performed and paid for by the company] had energy applications, including energy production, distribution, storage, and efficiency (excluding exploration and prospecting)? Example: Company B is a semiconductor manufacturer. Its products are not designed specifically for energy applications. In 2010, 10% of the domestic R&D performed by the company was focused on improving the energy efficiency of its products. Based on this, Company B reports "10%" for this question." and "Q4.25. What percentage of the amount reported in Questions 4–20 [domestic R&D performed by the company and paid for by others] had energy applications, including energy production, distribution, storage, and efficiency (excluding exploration and prospecting)?"

A report for the European Commission Joint Research Centre (Wiesenthal et al., 2009) found an even greater level of uncertainty in estimates of EU energy technology R&D, particularly given the 2008–2010 improvement in U.S. private R&D statistics. Energy R&D is funded by the EU, the member states, and by industry with both overlap and gaps in the available funding data sources. Although some EU records on the private sector are more comprehensive than those in the United States, Wiesenthal et al. resorted to collecting data from 136 large R&D performers from filings, corporate reports, and interviews. Their estimates include a total of €3.32 billion (roughly $4.5 billion in 2007) in EU-wide low-carbon energy R&D, 72% of which went to non-nuclear technologies; they estimate the funding split to be 11% by the EU, 33% by member countries, and 56% by industry (Wiesenthal et al., 2009). These estimates do not include any fossil fuel or energy efficiency R&D, so these figures are .not directly comparable to the U.S. estimates presented in Figure 4.4.

Venture capital (VC) investments are also a proxy measurement of energy innovation activity. Bloomberg New Energy Finance (BNEF) has been collecting information on VC and other investments (e.g., asset finance and private equity) in areas of "new energy"[4] (a large subset of energy innovation) since the year 2000. However, these data do not include R&D investments by corporations, and not all VC investments are dedicated to R&D. For comparison with NSF's estimated $5.3 billion of energy R&D investments by firms in 2007 in the United States and about $16 billion in 2008 according to the NSF BRDIS (2012 data are not yet available), the BNEF database reported $4 billion in new energy VC and private equity funding in 2007, $7 billion in 2008, and $4.3 billion in 2012 (Bloomberg New Energy Finance, 2013).

4.2.2. Underinvestment in Energy Technology Innovation

Even with limited data, experts broadly agree that U.S. investment in energy RD&D is inadequate (Nemet & Kammen, 2007; Dooley, 2010). This finding echoes longstanding economic evidence of general underinvestment in innovation. For decades, scholars have emphasized that innovation plays a key role in economic growth (Schumpeter, 1934; Solow, 1957). Because some benefits from innovation cannot be captured by individual firms – a fact which is known as innovation having spillovers – the private sector tends to underinvest in innovation, so policies are necessary to stimulate more investment (Arrow, 1962). Research and development require long lead times, and the outcomes are uncertain, so shareholder-focused firms tend to underinvest in RD&D (Anadon et al., 2009).

As a result of these market failures, social returns from investments in innovation are far higher than the returns that any single company is likely to earn. Government has

[4] Bloomberg New Energy Finance analysts track financial markets, filings, announcements, transactions and other indicators of activity in "clean energy," low carbon and carbon markets, including wind, solar, biomass, geothermal, efficiency, carbon capture and sequestration, carbon trading, services and support, etc. (Bloomberg, New Energy Finance, 2009).

a major role to play in stimulating socially optimal levels of innovation (Margolis & Kammen, 1999; Brown, 2001; Audretsch et al., 2002; Jaffe, Newell, & Stavins, 2005). To compound the problem, all energy-related issues are also subject to multiple market failures (Jaffe, 1998; Brown, 2001; Jaffe, Newell, & Stavins, 2005), as discussed in Chapter 1. In short, innovation in energy technologies suffers from all the usual troubles inherent in technology innovation and energy systems (Fri, 2003; Gallagher, Holdren, & Sagar, 2006).

The impact of these market failures is clear. In 2010 the United States spent an estimated $1.2 trillion on energy (EIA, 2012). According to the estimates in Figure 4.4, U.S. private energy R&D spending in 2010 was roughly $16 billion (NSF), which represents only 1.3% of energy expenditures in the U.S. economy. In contrast, in the total economy, R&D represented 2.8% of total GDP (NSF 2013) – energy R&D is at a lower intensity than the average over all industries. Similarly, Wiesenthal et al. (2009) estimate that U.S. energy R&D spending was equal to 1.1% of all U.S. R&D spending in 2007, compared to 2.9% in the EU and 15.2% of total R&D spending in Japan. Top firms in the U.S. information, computer, and health care industries each invest more in R&D than the federal government and private companies spend together on energy R&D (Battelle, 2010).

U.S. underinvestment in energy R&D is especially striking given the enormous energy-related environmental, security, and economic challenges that the United States and the world face. U.S. investment in energy RD&D is low by any measure: it is far smaller than total national energy expenditures or R&D in other industries, or energy R&D in other countries.

4.2.3. Funding and Performance of Energy R&D

To address market failures that undercut innovation, governments invest directly in activities that promote innovation, particularly in the early stages (Jaffe, 1998). The main federal agency that supports energy innovation is DOE, which also provides significant support for basic research in the physical and life sciences.[5] Although many DOE activities are unrelated to energy innovation,[6] the fraction that is dedicated to energy innovation represents a large share of the total U.S. activity in energy RD&D (roughly one fifth of total funding). As shown in Chapter 2, between 2000 and 2013 investments have gradually increased from $2.6 billion to $4.6 billion, with a $7.5 billion spike in 2009 attributable to the American Recovery and Reinvestment Act (ARRA) ($2010).[7] Although current levels of DOE energy RD&D investment are not

[5] Other agencies that support innovation in energy technologies include the Departments of Agriculture, Defense (including the Defense Advanced Research Projects Agency, or DARPA), and Interior, as well as the National Aeronautics and Space Administration (NASA).

[6] DOE's budget request for 2011 included $10 billion for Energy Programs and Science (which includes funding through the DOE Office of Science and the "applied offices" for EERE, FE, NE, and OE) compared to $11 billion for the National Nuclear Security Administration and $6 billion for clean-up activities (DOE, 2010).

[7] Other agencies beyond DOE are active in supporting energy RD&D. Using data from the Energy Innovation Tracker (2013) we found that other agencies invested about $1.2 billion in 2012 in energy RDD&D, which

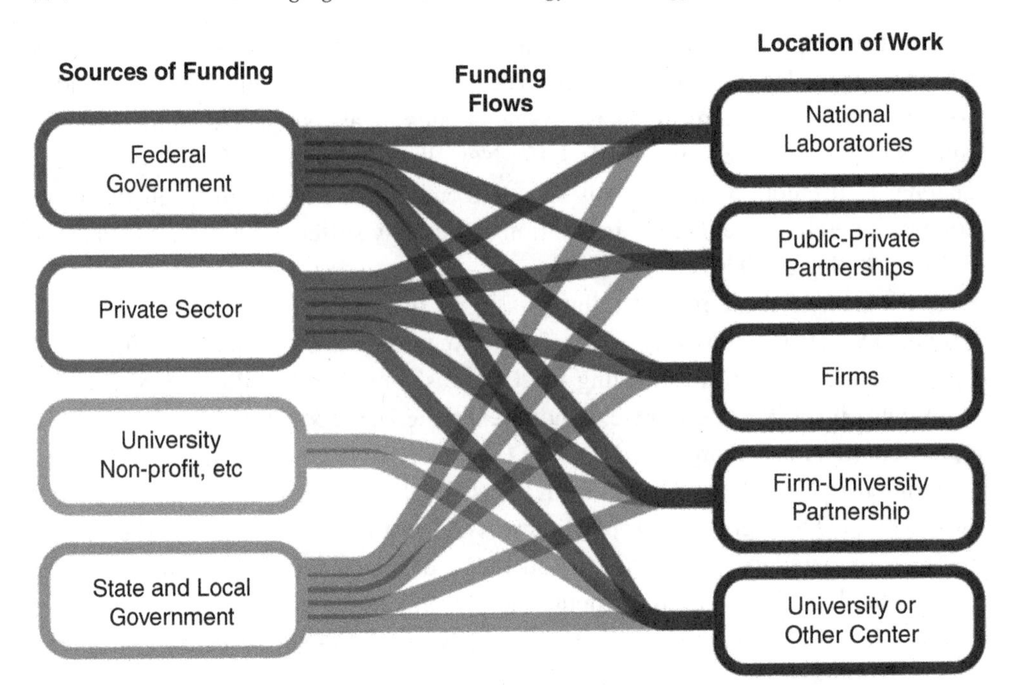

Figure 4.5. Funding for and location of work on energy RD&D. The federal government, the private sector, and state governments fund all locations (albeit to different extents).

close to the $8 billion invested in the late 1970s, this funding is very significant, more so when considering that DOE's goal is to use its activity to attract private-sector investment, to increase the value of private-sector efforts, and to push new technology into the marketplace (DOE, 2006).

Innovation is both a social and a technological enterprise. Systems for technological change cross organizational, field, and political boundaries and involve multiple interactions, feedbacks, and delays (Carlsson & Stankiewicz, 1991; Garud & Karnøe, 2003). In the energy innovation process, any party can conduct its own R&D, pay others to conduct it, conduct it on behalf of others, or induce others to fund or conduct their own. Figure 4.5 illustrates the web of funding pathways that supports this complex system.

The federal government distributes funding for energy RD&D to academic institutions, national laboratories, private firms, and partnerships, and provides facilities and services (primarily from the national labs) for research on behalf of other actors. Matching funding requirements attract additional resources from other parties, while research and experimentation tax credits make R&D more financially attractive for

includes (according to our interpretation of the data) direct spending for "basic research," "applied research," and "demonstrations" for the Departments of Commerce, Defense, Transportation, Agriculture, and the Environmental Protection Agency, NASA, and the NSF. Other Departments covered by the database that showed up as having no energy RD&D expenditures include Interior, Housing and Urban Development, and Treasury.

private companies. Private firms may fund energy R&D from revenue or capital, then conduct it themselves, outsource it to other firms or the national labs, or use the funds to sponsor academic research. Many of the flows depicted in Figure 4.5 have yet to be quantified.[8]

In spite of the significant improvements in data collection by the NSF BRDIS shown in Figure 4.4, the United States needs much clearer and more detailed information about energy innovation in the private sector and public–private partnerships. Policymakers are attempting to correct market failures and to encourage private-sector innovation without knowing what activities are currently taking place or what is needed in what technology arena, and without accurate information about what types of projects produce the best results under different situations. This chapter attempts to answer some of those questions.

As described in detail in Anadon et al. (2011), we used several types of information to gain insight into public and private-sector energy innovation and how energy R&D efforts could be made more effective. In the remainder of the chapter we present the results from a new database constructed from public data on DOE funding for energy innovation support outside of the DOE, and a survey we conducted of business establishments to improve information about private-sector activities. Managerial and strategic lessons are drawn from interviews with experts, prior studies, and the aforementioned survey of business establishments.

4.3. Private Sector Activities in Energy Innovation

4.3.1. Energy Innovation Landscape

There is no trade association, industrial classification or code, or product category that defines a coherent U.S. "energy industry." Creating such a code would leave out important energy activities, while inevitably including too many non-energy-related activities. Without a proper definition, it is impossible to estimate the energy industry's R&D expenditures using existing industry codes. The NSF BRDIS represents an important step, but given its scope it cannot include questions about the technology area of focus, startups, the impact of different policies, and the factors that shape R&D

[8] For completeness we include here what values we could estimate, calculated by a variety of measures, including (1) budget expended for the U.S. National Laboratories by DOE applied energy offices in 2008 (from FY09 DOE budget justification): $1.7 billion; (2) DOE assistance in energy RD&D in 2008 mainly through grants and cooperative agreements (from USA Spending database): $836 million overall ($797 million of which was dedicated to cost-shared projects and $38 million of which was dedicated to projects solely funded by the federal government). DOE assistance funding for energy RD&D had the following breakdown by recipient (figures may not sum due to rounding); (3) $529 million with for-profit recipients, $512 million of which went to cost-shared projects; (4) $211 million to universities, $194 million of which went to cost-shared projects; (5) $36 million to state, local, and tribal governments; $33 million of which went to cost-shared projects; (6) $61 million to non-profits and others, $57 million of which went to cost-shared projects; (7) non-federal contributions to cost-shared projects in the USA Spending database were $771 million in 2008: $89 million to projects at universities, $43 million to projects in governments, and $610 million to projects in the private sector; and (8) payments to the National Laboratories for work from firms were worth $308 million in 2009 (Farris, 2010).

decisions. It is also impossible to determine R&D expenditures by industry related to energy, without significant caveats. Instead, we have undertaken to determine which firms undertake activities in (a) energy and (b) innovation. Then we examine the perspectives and spending decisions of establishments that are engaged in energy-related innovation.

To identify and describe entities involved in energy technology innovation, we developed a classification from the NAICS code and SIC code definitions and our own expert interviews, which included both producers and consumers of energy technologies, including:

- *Firms in energy resource extraction, conversion, or distribution.* Some firms extract fuels and convert, develop, or distribute energy. This segment includes mining companies, utilities companies, refineries, independent power producers, and biofuel producers. These actors might use or develop new or improved energy equipment, and they are the customers for energy equipment.
- *Firms supplying energy equipment.* Some firms supply energy equipment, such as turbines, boilers, solar panels, transformers, etc. Others provide specialized construction, such as for power lines, or provide equipment to manufacture energy components, such as machinery for manufacturing photovoltaic cells. Energy technology is their core business, and innovation helps energy consumers or other energy companies.
- *Firms that supply resources to energy equipment or services firms.* Upstream of the energy industry are suppliers of capital goods and important components and materials. Though energy might be a small part of their business, innovation could be important if it enables improved energy equipment or services.

The next three categories denote businesses that are outside of energy services or equipment but have a large stake in energy innovation:

- *Firms that use large amounts of energy in their businesses, such as aluminum manufacturers.* Energy-intensive businesses may attempt to make their processes more energy efficient, and we want to capture whatever fraction of those efforts is truly innovative.
- *Firms manufacturing products that are energy intensive, such as automobile manufacturers or air conditioners.* Manufacturers of products that use significant energy develop new products and improve existing lines, and some fraction of that effort represents energy innovation.
- *Firms producing equipment that affects the energy efficiency of buildings.* Other firms manufacture equipment that does not use energy or that is not energy intensive to manufacture, but that affects how energy is used, for example, insulation, high-efficiency windows, thermostats, and other controls. These firms' innovative efforts make energy reduction more affordable.

Finally, two types of actors provide services for other firms' energy technology. As a measurement issue, the activities of these actors could be found either by asking the clients how much they outsource, or by asking the providers how much energy-related business they conduct; care must be taken in statistical estimates to prevent under- or over-counting. These two types of firms are listed below:

- *Firms providing engineering or R&D services.* Engineering or R&D service providers might support a variety of customers or work on a variety of projects, some of which should count as energy innovation.
- *Consultants and energy service companies (ESCOs).* Some consultants and energy service companies help customers reduce or manage energy use. These firms might innovate to provide better solutions for their clients.

The diversity of the types of businesses potentially engaged in energy innovation complicates measurement of energy innovation. In addition, as we mentioned above, about one quarter of all R&D performing firms are engaged in energy-related R&D, as shown by the 2010 NSF BRDIS. We interpret this finding to indicate that energy innovation is more prevalent among R&D performers than previously thought within most industry groups.

We also define innovation very broadly. Considering energy innovation activities beyond traditional R&D results in a more diverse set of actors, and what they recognize and how they describe innovation varies as well. Based on our survey findings, energy innovation can mean any of the following for business leaders:

- *Basic and applied research.* Discovering new knowledge through answering scientific and technical questions related to new technology.
- *Technology development.* Applying knowledge for new or significantly improved products and services – although there is some overlap, this generally refers to new technology as opposed to new products in an existing technology.
- *Pilots or field demonstrations.* Pre-commercial-scale or nearly commercial-scale experiments to gain experience, test the technology, or prove the viability of the product or methods of making the product.
- *Initial commercial-scale activities.* Either the first product or first facility to manufacture the product; there is an experimental aspect but the activity is expected to transition to normal commercial operations.
- *Product design or improvement.* Sometimes this is the last phase of technology development, but can also apply to later iterations of product lines under existing technology, or products in existing lines.
- *Process design or improvement.* Innovation aimed at improving how current products and services are created, or at enabling the creation of improved products or services. Efficiency and quality initiatives are important examples.
- *Technology search or evaluation.* Finding or adapting technology solutions from other fields or industries to enable innovation in one's own field. Our interviews revealed this to be especially important to startup and venture capital firms.
- *Intellectual property management.* Obtaining and selling licenses for patented technology, to capitalize on or to enable one's own firm's innovation.

4.3.2. Estimates of Private Sector Energy Innovation

The *Survey of Energy Innovation* consisted of two questionnaires, administered in stages. The first questionnaire identified establishments involved in energy and innovation work, and the second collected details from firms that identified themselves as conducting RD&D on energy systems and technologies. A separate telephone questionnaire was administered to a sample of nonrespondents to estimate bias. Of the 192 respondents to the screener questionnaire, 107 (56%) reported some level of involvement with energy, as described in Section 4.4.1; 88 (46%) reported innovative activities or use of the results or innovation in other locations; and 54 (28%) reported that their innovation activities and energy activities are related.[9] The establishments of greatest interest are those that actually performed, funded, or directed work on innovation. Of the firms surveyed, 61 (32%) conducted innovation themselves, and 38 of these (20% of the total) conducted innovation in energy.[10] Our findings indicate that about a fifth of business establishments are involved in energy and about a fifth in innovation.[11] Of the 192 respondents, 33 described themselves as startups. Establishments that reported innovation activities related to energy activities were sent a follow-up survey instrument.

Assuming that none of the nonrespondents are involved in energy or innovation, our survey still shows that 3.4% of all establishments surveyed are involved in energy, and 2.4% are involved in energy-related innovation (with an approximate error of 1.3%). This adds up to about 260,000 business establishments conducting energy-related R&D in the United States – the full universe for the stratified survey included 10.8 million establishments – making energy R&D a crucial activity for a very large number of businesses. In terms of firms, about 0.57% of U.S. firms performed and paid for energy-related R&D according to the NSF BRDIS (11,500/2.01 million).[12] Table 4.1 presents the detailed estimates from the survey with standard margins of error,[13] and the main findings are presented in the following.

- *Energy technology innovation is widespread.* Based on the weighted sample, 17% of establishments are involved in both energy and innovation, with 16% specifically involved in energy innovation. About a tenth of establishments actually perform, fund, or direct energy

[9] Firms could answer "yes" to more than one of these questions.

[10] The 82 establishments that did not answer the screening questionnaire but were interviewed thereafter were slightly less likely to engage in energy innovation, but the differences were not statistically discernable: 45% reported energy-related activities, 22% conduct innovation, and 16% performed innovation related to energy.

[11] These are likely to be overestimates owing to response bias. In an effort to measure response bias, we contacted 258 of the nonrespondents. Eighty-two of these completed a telephone interview; 37 (45%) reported being involved in energy and 13 (16%) in energy-related innovation. Accounting for the sample design, an estimated 9.9% of the nonrespondents are involved in energy and 4.3% in energy-related innovation – less than the estimate from those completing the survey proper, and evidence of response bias. Further evidence of response bias is that the weighted estimates of energy/energy-related innovation are 28.6%/24.0% among those answering the first survey, and only 8.3%/2.6% from the follow-up survey.

[12] The number of firms in the NSF survey (1.35 million) is different from the number of business establishments in the Harvard survey (10.8 million) primarily because one firm can have many business establishments.

[13] Estimates are calculated from 180 respondents. Standard errors are calculated using the multifactor jackknife method and 95% confidence interval.

Table 4.1. *Prevalence estimates from the Harvard Survey of Energy Innovation in U.S. Businesses*

Activity	Estimate (%)	Standard error (%)
Involved in energy	21.0	4.6
Services, goods, equipment	15.8	10.7
Energy use	9.8	7.1
Innovation providers	5.4	4.6
Involved in innovation	19.1	10.7
Perform, fund, direct	12.7	10.2
Use of others	18.8	10.7
Both energy and innovation	17.3	10.6
Involved in energy-related innovation	16.0	10.6
Conducting energy-related innovation	11.1	10.3
Startup companies	0.64	0.3
Involved in energy	0.57	0.3
In energy-related innovation	0.27	0,2
Floor estimate of establishments involved in energy	3.4	1.1
Floor estimate of establishments in energy-related innovation	2.4	1.3

innovation and another 5% consider energy innovation by others as important to their business.[14] Thus, energy innovation seems far more common than prior estimates of energy R&D.

- *Startup companies are more engaged than established companies in energy innovation.* Over the entire population, just over six-tenths of a percent of all establishments are startups, of which nearly half are engaged in energy technology innovation. Thirty-three respondents reported $394 million in capital raised to date. Twenty of these were both startups and energy innovation participants, representing most ($330 million) of the capital raised to date. The sample yields an estimate of 29,000 (standard error, or "se" = 20,000) startups engaged in energy innovation, with $31 billion (se = $30 billion) in capital raised to date.
- *Energy innovation includes both normal business and special projects.* Most establishments (17) report conducting energy innovation as a normal part of business. Eight conduct energy innovation only as special or one-off projects, and four use only energy innovation developed by others. More startups conduct energy innovation as a normal part business (12/15) than do established firms (5/14).
- *Small-scale energy innovation is important.* The 17 establishments that conduct energy innovation as a normal part of business report $33 million (± $21 million) in innovation expenses in 2009. Most of these businesses (11/17) report less than $200,000; the total value is dominated by two startups in the $5 to $20 million range. In the special-project only businesses (8/17), the most recent projects were reported as totaling $5.4 million for an average cost of $768,000 per project.

[14] Although these are likely to be overestimates, a floor on the actual value of establishments involved in energy innovation is between 1.1% and 3.7%.

- *Private sector energy innovation spans all stages of research and development.* Even our small sample of only 17 establishments represents all stages of innovation and all parts of the energy innovation landscape. Every establishment engages in more than one stage of innovation; technology development (15) and pilots or field demonstrations (14) are the most commonly reported. Innovation for products to reduce customers' energy use (9) and to provide technical or R&D services to clients (7) are the most common purposes for innovation. Of the special projects, three had purposes related to saving energy, three mentioned developing a product of some kind, and two were related to processes. Most energy technologies are found even in a small sample. Among the 29 companies, 70 technology areas were mentioned in which they are involved with innovation: solar is the most common (11); 17 of the technology areas covered different projects related to increasing the energy efficiency of various products; and wind, electric generation, industrial processes, and energy storage were each mentioned by six establishments.
- *Additional incentives for energy innovation are needed.* Given the many market failures impeding private sector investment in energy innovation, additional incentives are justified, if they could cost-effectively leverage additional private sector spending. Our survey indicated that government cost-sharing and deployment-related incentives (such as production tax credits) are generally more important in private sector innovation investment decisions than R&D tax credits, but further research is needed to design the optimal package of incentives for private sector energy innovation.

4.3.3. Process of Private Sector Energy Innovation

The follow-up survey collected information on how businesses that are active in energy technology innovation made decisions, and what factors promoted or hindered energy innovation in their industry. In addition to the source of funds and the location of work, respondents were asked to rate the three most important factors in a number of areas: information, sources of innovation, related fields, policy and market factors, and trends and barriers. Their responses yielded the following insights:

- *Most funding and performance is internal.* Fifty-three percent of reported funding was provided by the reporting establishment itself, followed by federal grants and contracts (24%). Eighty-six percent of innovation work was performed at the reporting establishment, with 10% outsourced to other U.S. companies. Only 0.1% of company-funded research was done at national laboratories.
- *Short payback times dominate decision making.* Whether for normal business activities or for special projects, firms use similar criteria to evaluate the financial impact of energy innovation investments. Most common (10 out of 24) was estimating a period to recoup the cost of innovation, with the average time being 2.8 years. Nine out of the 24 reported that they do not formally measure financial impact.
- *Information that businesses use to make energy innovation decisions comes from a variety of sources.* Institutional sources such as government and academia are used by 17 of the 24 businesses that specified. Twenty respondents mentioned private sector sources, but these include internal sources, other companies up or down the supply chain, and competitors.

- *Energy prices drive innovation more than opportunities.* Energy prices were mentioned by 9 out of the 24 as a driver of energy innovation decisions and were the most-mentioned factor for promoting innovation by 16 respondents, followed by cost-cutting opportunities (11). There are 19 mentions of various market or market creation policies, but there is no dominant, single market factor.
- *Grants have more impact than tax credits.* Policies that directly fund innovation are mentioned 20 times as factors for promoting innovation, with government R&D grants or support (9) being the most important of these.
- *The innovation process and the lack of markets are equal barriers.* The single most challenging factor for innovation is high commercialization cost. If the types of barriers faced by businesses are grouped by increasing level of challenge, they are: difficulties associated with the innovation process (26), market barriers (25), state of science/technology issues (12), regulatory barriers (6), and intellectual property issues (2).
- *Cooperation is important.* The most beneficial trends mentioned by respondents were cooperation with government (12), universities (10), and with other companies (6), and in mergers and acquisitions (11).
- *Most important innovations come from within the industry.* Both respondents' own firms (18) and other firms in the same industry (20) are listed as the most important sources of innovation. National laboratories (12), universities (11), and other industries (10) are all frequently cited as the second or third most important origins of innovation. Electrical and electronic engineering (13) and materials science (12) produce innovations that have the most positive impact.
- *Innovation is central to startup businesses.* The 15 startup respondents report, on average, that 75% of their startup capital has been spent on energy innovation. These respondents raised $284 million in capital to date. Twelve were using founder capital for operations in 2009, including five for whom it was the only source of funding; the next most common source (5 out of 15) was government grants or contracts. Thirteen of the 15 own some amount of intellectual property, and 8 of the 13 describe their intellectual property as being generated after startup.

4.3.4. Reflections on Private Sector Energy Innovation

Even though respondents are more involved in energy technology innovation than the general population of business establishments,[15] this study and the recent results from the NSF BRDIS provide evidence that energy innovation is important to far more businesses than prior studies have found, and that it is conducted by more companies than prior estimates. Even though most of the investment comes from the largest

[15] With the response rate of 14.7% there is likely a difference between those who did and those who did not answer the survey. Among those who answered after the first mailing, 37 of 107 (35%) reported being involved in energy innovation. Of responders to the second mailing – who may be more similar to the nonresponders – only 17/85 (20%) reported being involved in energy innovation. In follow-up calls to judge the scale of non-response bias, 13/82 (16%) reported energy innovation. There is possible response bias even in the nonrespondent bias study, so we assume all establishments we failed to reach would have answered "no" to the energy and innovation questions in order to find the floor for energy innovation. Even this extreme assumption gives an estimate for involvement in energy innovation of 2.4 ($\pm$1.3)%, or at least 120,000 establishments. If being involved in energy implies potential benefit from energy technology, the floor on the impact of energy innovation is 3.4 ($\pm$1.1)%.

volume businesses, most of the firms engaged in energy technology innovation (ETI) are making small investments that have not been previously captured. Firms that only conduct special projects in energy innovation might be missed by a more narrow focus on R&D providers. The broad impact of energy technology indicates that it should be a priority for federal resources.

In addition, the respondents' answers provide insight into how businesses that engage in ETI operate. Markets and market policies can drive innovation, but it is the lack of markets, lack of market policies, and the difficulty of bringing products to market that are the most significant barriers. Energy prices and cost-cutting opportunities are the largest drivers of innovation. Based on business observations, a mechanism to price carbon should encourage innovation by setting a price signal and creating a market. Responses to our survey suggest that the most significant roles government agencies can play are as sources of information and partners for collaboration. Government also can create policies to ensure that the full costs and benefits of energy technologies are priced in the market, and can create grants and contracts targeted for different types of technologies, firms, and projects.

4.4. DOE and Partnership Activities in Energy Innovation

4.4.1. Types of DOE Activities

The DOE receives appropriations from Congress but also collects funds from industry and enters into multiyear agreements, so it is difficult to quantify how much the agency invests in energy technology development in any given year. As discussed in Chapter 2, in the FY 2010 budget, $5.3 billion was appropriated for energy technology and energy-related science, out of a total appropriation of $26.6 billion (DOE, 2010; Gallagher & Anadon, 2010). In addition the department received $3.7 billion for energy RD&D and Basic Energy Sciences in one-time economic stimulus funds under the American Recovery and Reinvestment Act, including sciences. Without including stimulus funds, the investment in energy technology for 2010 would fall well below the inflation-adjusted levels of the late 1970s of about $8 billion (2010$). Investments in 2012 grew to $4.6 billion, which is still below the oil crisis levels.

DOE conducts research through the national laboratories, awards grants and contracts to conduct research, and conducts R&D projects with partners. It also thrives to introduce technologies into the market and provides funding and expertise to state agencies and local governments. Although there is some overlap, funding to the private sector falls under activities that the DOE refers to as "R&D Support" (transfer of funds to a non-federal party) and "Technology Transfer" (mechanisms for transferring federal intellectual property to a non-federal party) (DOE, 2009c). Either type can originate at the national labs, Operations Offices, DOE headquarters, or through the initiative of the non-federal partner. Most activities solicit applications at some point,

Table 4.2. *Roles of Federal entities in energy RD&D support*

Office or level (example)	Roles
Congress	Authorize programs, appropriate funds, oversight
Administration	Direction, strategy, budget
Office of Management and Budget	Accumulate and evaluate budget requests across departments
Secretary of Energy	Direction, strategy, budget
Office of Policy and Intl. Affairs	Analysis to support budget and program decisions
Office of Management	Policy, data collection, oversight
Chief Financial Officer	Assemble DOE budget requests, fiscal oversight
Inspector General	Reviews, oversight
Office of Scientific and Technical Information	Data collection
Program Offices (EERE)	Write FOAs, select recipients, planning, budget input
Programs (Solar)	Technical expertise, FOA and planning input
Operations Offices (Golden)	Issue FOAs, manage awards
National Laboratory	Negotiate CRADAs, WFOs, UF, licensing; technical expertise
ARPA-E	Merge of program office, program, and operations office
Office of Science	Program office plus all SBIR/STTR
NETL	Both a NL and ops office plus all unsolicited proposals

Source: descriptions throughout the DOE website (www.energy.gov).

and unsolicited proposals may also be accepted. Table 4.2 summarizes the complex network of relationships through which DOE funds, manages, and monitors these activities. Because each type of agreement has its own set of procedures, there exists no central record of partnerships collecting information about the technology covered or the results.

Although over the past 10 years or so the Office of Energy Efficiency and Renewable Energy (EERE) has been very active developing tools to learn from their previous experience,[16] in general the information available for activities across all DOE Offices is insufficient, particularly related to the technologies, type of work, and main findings and results from the work supported. As we discuss in the conclusions, this is a major

[16] EERE has provided overview, planning, implementation, and results documents for all of its programs (http://www1.eere.energy.gov/pir/table.html). It has also been working on conducting retrospective benefits estimation (http://www1.eere.energy.gov/ba/pba/program_evaluation/index.html). It has also started to conduct Annual Merit Review & Peer Evaluation Meetings to evaluate whether activities (mostly those funded by competitive solicitations) are being effective using peer review. These meetings result in the preparation of evaluation documents of work "in-progress," which are also available online. This link http://www.hydrogen.energy.gov/annual_review.html provides information on the Annual Merit Review & Evaluation meeting for the Hydrogen and Fuel Cells Program. Finally, EERE has also conducted and published case studies on their Small Business Innovation Research (SBIR) work. See: http://www1.eere.energy.gov/hydrogenandfuelcells/pdfs/sbir_proton.pdf for an example. More broadly, EERE has an effort underway to link SBIR companies with the Clean Energy Alliance of Business Incubators to provide mentoring and other support to these companies. This effort includes regular monitoring by a third party to evaluate the effectiveness of this partnership.

Table 4.3. *Types of R&D Support and technology transfer mechanisms at the DOE*

Types of DOE interactions with non-DOE entitles	Subcategory	Funding	Work
R&D support	Cooperative agreement	Shared	Shared
	R&D consortia	Shared	Shared
	Grant	DOE[a]	Firm
	Contract	Shared[a]	Firm
Technology transfer	CRADA	Firm	Shared
	Work for Others	Firm	DOE
	User Facility	Firm	Firm
	IP agreements	Firm	DOE
	Personnel exchanges	None	Host
	VC initiatives	Firm	Either

[a] Grants are government-funded, but may include provisions requiring matching funds; contracts with R&D support are usually cost-shared, but wholly government-funded contracts are typical in other DOE activities, such as facilities management.

area for improvement that would allow researchers and analysis outside of DOE to generate further opportunities for improvement. The only dataset that we were able to find capturing some aspects of funding from DOE to external actors came from www.usaspending.gov, a public disclosure project that resulted from the Federal Funding Accountability and Transparency Act of 2006 (S. 2590) (Transparency Act, 2006), which mandates disclosure of contracts and assistance by the White House Office of Management Budget (OMB). Given that the goal of the dataset was not to provide data to improve the management of energy RD&D partnerships over time, this dataset does not contain information about the type of technology, project, or outcome of the funding agreements.

The DOE supports R&D activities through financial awards (also known as grants), contracts, cooperative agreements, management and operations contracts, personnel exchange programs, R&D consortia, and Cooperative Research and Development Agreements (CRADAs) (DOE, 2009c). Technology transfer activities include intellectual property (IP) agreements, CRADAs, licensing, Work-for-Others agreements (WFOs), user facility (UF) agreements, technical consulting, and personnel exchanges (DOE, 2007). Recent additions to these mechanisms include pilot programs involving venture capital firms (DOE, 2007). Table 4.3 describes key characteristics of these agreements.

Federal agencies use varying definitions for R&D support mechanisms: they may call these initiatives “assistance” or “awards” in different data sources, and these programs often overlap with technology transfer mechanisms. DOE uses, cooperative agreements to fund projects either completely or partially. R&D Consortia involve multiple federal and non-federal partners in cost- and work-sharing agreements. Major initiatives of this type (e.g., FutureGen, FreedomCar) are structured as multiple cooperative agreements. In such agreements, grants are used to transfer funds to the non-federal party on a competitive basis (project grants), or through programs that distribute funds based on population or other measures (formula grants). In some cases, grants require the grantee to secure additional funding to be eligible for the grant. Occasionally, assistance may come in the form of equipment, service, direct payment, or other forms. Finally, the DOE may issue a procurement contract to commission research by another party. Typically these projects are required to be cost shared and mutually beneficial.

Technology transfer activities at the DOE are controlled by national laboratories and facilities and do not transfer funds to the non-federal party. For example, CRADAs involve shared personnel, services, equipment, or facilities, as well as shared results, and the cost of government work is usually covered by the non-federal partner. In WFO agreements, the government conducts work on behalf of another party, while in UF agreements, the party conducts its own work using government equipment. Personnel exchanges allow temporary transfer of researchers either to or from a national laboratory to share knowledge. Finally, all types of intellectual property transactions – patents, copyright, sales, licensing, or negotiated agreements – are considered to be technology transfer activities.

DOE's Small Business Innovation Research (SBIR) and Small Business Technology Transfer (STTR) grants involve both support and transfer. They are similar programs, with STTR grants requiring significant participation by a university or other research partner. All federal agencies conducting significant R&D must set aside 2.8% of funding for these programs. Phase I grants are meant to be feasibility studies, and are usually 9-month activities awarded $100,000. If successful, Phase II grants can be awarded up to $750,000 over two years. Awardees often receive technical assistance or advice from DOE researchers (DOE, 2009d).

All of these mechanisms involve interaction between the DOE or national lab researchers and researchers at universities or private firms, but only a minority of agreements require DOE and university or private sector researchers to share risk, resources, and decision-making authority – the elements that we believe define a true partnership. CRADAs, cooperative agreements, and R&D Consortia are structured to allow partnerships as we define them. Of these, only CRADAs and cooperative agreements are found in the USASpending.gov data, which means that R&D Consortia are not represented in the data presented in Section 4.4.2.

4.4.2. Trends in DOE Activities

DOE's 21,326 awards for assistance that were active between FY 2000 and 2009 included 5,648 energy RD&D projects (funded by the DOE applied offices) and 10,601 science projects (funded by the DOE Office of Science). During this period, the DOE invested $7.2 billion in energy RD&D assistance (60% of which was directed

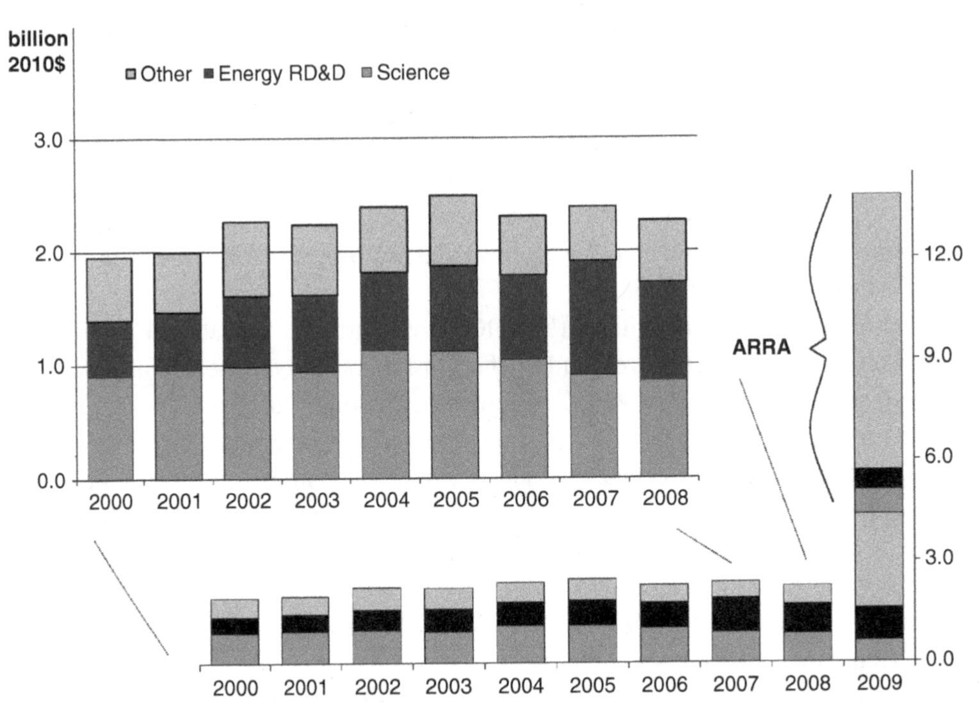

Figure 4.6. Funding for assistance, mostly in the form of cooperative agreements and grants, to private firms, academic institutions, and other institutions, by the DOE in major categories (energy RD&D, Science, and "other"). The fluctuation in spending through fiscal year 2008 on all three categories of assistance is apparent when examined; the scale is expanded to show the spending under ARRA.

to the private sector) and $9.1 billion in Science (74% of which was directed to academic institutions). These totals include funding for SBIR and STTR programs, which are set at a fixed percentage of the R&D investments. Because funding for these programs is a constant fraction of total DOE investments – which are less volatile than funding for individual technology areas or programs – SBIR and STTR funding is less volatile than that for other programs (DOE, 2009d). As a whole, assistance comprises a large segment of the DOE budget for energy innovation, fluctuating at about 55% for Science and 30% for energy RD&D.

Figure 4.6 shows how the assistance portion of the DOE's budget has been spent on energy RD&D, Science, and "other" programs unrelated to energy RD&D, including spending under ARRA. The Recovery Act increased all types of assistance. From 2000 to 2008, spending on total science plus energy RD&D increased on average 3% per year in real dollars, but that trend is almost lost amid yearly budget fluctuations. Over the same period, the mix of spending has generally shifted away from science toward the energy RD&D programs – from 65/35 in favor of Science to 53/47 in favor of energy RD&D.

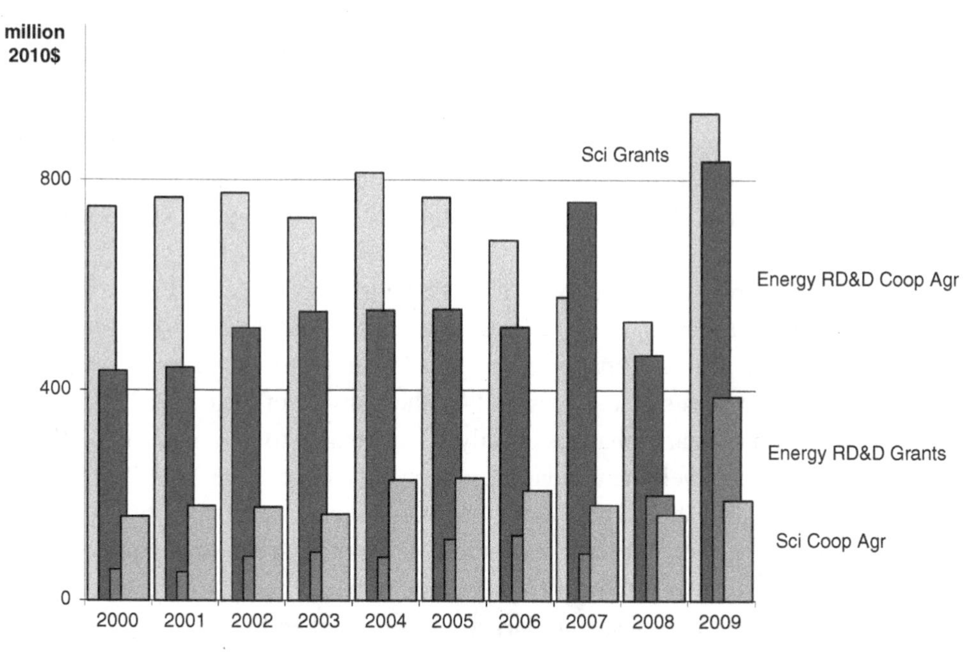

Figure 4.7. Funding for DOE grants and cooperative agreements in Science and energy RD&D.

Volatility is apparent in many of the spending categories. Narayanamurti, Anadon and Sagar (2009) calculate the volatility in budgets by the standard deviation in year-to-year changes, which averages 27% for major technology categories (e.g., coal, petroleum, gas, transportation, industry, and buildings), indicating that in about one out of three fiscal years, the budget for a particular technology program changed by at least 27%. Funding for energy RD&D and Science in the form of cooperative agreements and grants as a whole is less volatile than that for coal, petroleum, gas, transportation industry, and buildings. Standard deviation in year-to-year spending changes 6% for all assistance, 9% in Science, and 14% in energy RD&D. In specific categories detailed later, however, volatility is much higher, as moderate changes in overall funding multiply by larger changes in priority (in this case, and in cases below, the large funding increases in fiscal year 2009 are not included in volatility calculations).

Two main mechanisms – grants and cooperative agreements – make up more than 99% of assistance by the DOE, whether by funding or number of agreements. Grants may come with technical advice, but are, in general, simple transfers of funds to the recipient. Cooperative agreements allow for much closer partnerships. Figure 4.7 shows trends in these types of agreements, as reported in USA Spending. Between 2000 and 2009 the Office of Science consistently awarded around 15 times as many grants as cooperative agreements (between 2,400 and 3,000 grants per year vs. 100 to

400 cooperative agreements per year), whereas the energy RD&D programs award around 1.5 times as many cooperative agreements as grants (between 1,000 and 1,500 cooperative agreement per year vs. 700 to 1,000 grants per year). Science awards 2.5 times as many projects as energy RD&D, but the average project is far smaller. Science grants are usually the largest part of the four funding categories shown in Figure 4.3 (except in 2007), but only by about 28% over energy RD&D cooperative agreements.

Two trends are apparent. First, the Office of Science encourages research through a large number of relatively small, arms-length transactions, while the offices supporting energy RD&D use far more partnerships and larger projects. Second, the energy RD&D effort is far less stable. Measured by the number of awards, Science has a standard deviation of yearly changes of only 3% compared to 17% in energy RD&D. Measured by funding levels the standard deviation of yearly changes is 13% in Science and 34% in energy RD&D. These large variations are not explained by any stated DOE strategies. The fact that basic research is less controversial than applied R&D as a role for government (Merrill, 1984) is likely to be at least partly responsible. In addition, energy RD&D projects are larger on average than Science projects, so the start and end of particular projects contributes to more volatility. Whatever the cause, these fluctuations make it difficult for both the government and its partners to manage projects effectively.

Science and energy RD&D also have quite different patterns of recipients and cost-sharing. The Office of Science awards 74% of its funding to colleges and universities and 13% to small businesses. Almost all Science awards (97%) go to projects funded entirely by DOE. The distribution of cost-sharing and partners within energy RD&D offices is more complex. As Figure 4.8 shows, larger businesses receive nearly half of all energy RD&D assistance, on average. Higher education receives the next highest amount of assistance, followed by small businesses and non-profits. This distribution is once again volatile, with an average standard deviation of yearly changes of 26% (counting neither the large change in 2009 nor the small and variable "other" category).

Most energy RD&D projects involve some cost-sharing, which can be measured in several ways. Table 4.4 shows the number of projects that include DOE and partner contributions in a particular year and over the duration of the project, along with the allocation of funding. The percent of projects that are shared each year ranged from 60% in 2000 to 28% in 2009.[17] DOE contributes just under half of the total funding over the time period under consideration, with an especially low contribution in 2006 (an abnormality due to a $2 billion private contribution to a coal plant – the Mesaba Energy Project in Minnesota, in which MEP LLC invested $2 billion and

[17] Note that the total percentage of shared projects – 69% – is greater than that in any other year because each party may contribute in different years to a multiyear project. For example, if the DOE contributes a million dollars in 2002 and a firm contributes a million dollars in 2003 to the same project, the stats will show "no sharing" in both 2002 and 2003, since neither year had both federal and non-federal funding.

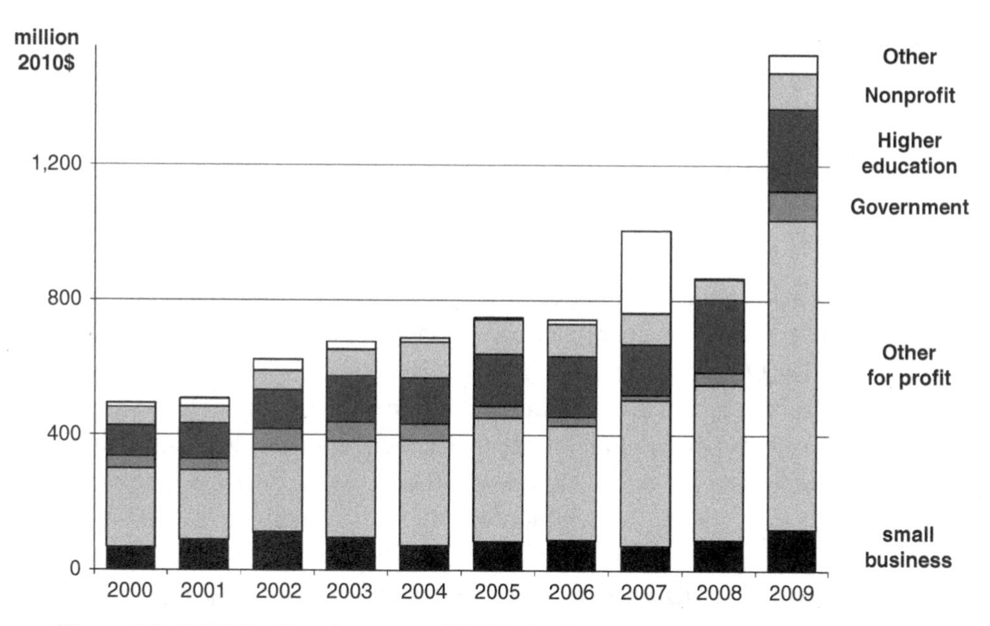

Figure 4.8. DOE funding for energy RD&D funding by type of recipient. The vast majority of these projects are funded by grants or cooperative agreements.

DOE $22 million for research at the plant). Overall, non-federal partners contributed $9.7 billion to projects from 2000 through 2009, $9.5 billion of which was in energy RD&D.

For most years, the largest part of funding goes to projects where DOE contributes between 40% and 80% of the cost of the project. In 2007, DOE allocated 41% of its

Table 4.4. *Cost-shared projects and level of cost-sharing in DOE assistance (mostly grants and cooperative agreements) for energy RD&D*

Year	Total active projects	Active cost-shared projects	Percent of shared projects	Contribution of federal funds to yearly funding of projects
2000	866	519	60	44
2001	884	467	53	54
2002	1.218	669	55	55
2003	1,289	626	49	55
2004	1.220	648	53	44
2005	1,296	603	47	41
2006	1,018	345	34	20
2007	879	291	33	56
2008	960	445	46	53
2009	1.121	316	28	46
Total (2000–2009)	5.648	3,888	69	43

funding for energy RD&D assistance to projects where a partner contributed at least 60% of the total, while in 2009, 41% of funding went to projects that are (thus far, see footnote 7) solely federally funded. The average standard deviation of yearly changes across cost-sharing categories is 50%.

4.4.3. Reflections on Trends in DOE Activities

Details concerning grants and cooperative agreements are important because they represent about 30% of the energy RD&D budget and 55% of the Office of Science (OS) budget. The OS cooperates mostly with universities – two thirds of the recipients and 74% of the funding are in places of higher education. Energy RD&D cooperation is spread more evenly – 15% of recipients are local government, 36% universities, and 38% firms. But firms received $4.3 billion or 60% of the energy RD&D collaboration funding, while contributing $6.5 billion or 39% of the total energy RD&D effort. Yet, it is difficult to find data on these agreements beyond their number and the total amount of money that DOE has invested in them in a comprehensive manner across DOE Offices, particularly outside of EERE. Previous studies provide additional evidence that information gathering and dissemination is an area of improvement for DOE. A 2005 report from the DOE Office of the Inspector General found that for many cooperative agreements in EERE, "project managers either had not completed or did not document reviews to evaluate the merit and technical aspects of the project" (DOE 2005). In A 2009 Government Accountability Office report, for example, found that "DOE cannot determine its laboratories' effectiveness in transferring technologies outside of DOE because it has not yet established department wide goals for technology transfer and lacks reliable performance data" (GAO 2009).

Information that is not publicly available for all activities with the private sector across DOE (with the exception of EERE efforts described in Section 4.4.1) includes the design of agreements, the outcomes of partnerships, the technology areas supported (what funds were directed toward entities working on biomass, solar power, buildings, etc.), and the lessons learned. Without this type of information strategic planners thinking about the role of DOE as a player in all major energy technologies cannot have been aware of the allocation of funding for projects and project types among technologies, technological stages (e.g., basic research, applied research, pilot, or demonstration stage), or partners. It would be difficult to conclude that partnerships are an element of strategic decisions if the DOE does not have an organized method for understanding its complete portfolio in a standardized fashion across its many different programs. For R&D managers, there does not seem to be a way that project design can take into account lessons from past projects from across the department (again, this type of review seems to be more institutionalized in EERE), other than the personal experience of the researchers involved. Though the circumstances of each partnership differ to some extent, project managers and decision makers would benefit from information about to learn what elements of partnership agreements are associated with projects that were/are useful and what elements are associated with projects that were/are not.

Over the past decade, DOE's total spending for project assistance has increased – especially in energy RD&D. ARRA funds boosted all types of DOE assistance, although they remain below the levels seen in the late 1970s. Though the broad trends are promising, funding volatility remains high, particularly for specific programs, and volatility in cost-sharing is significant. This volatility undercuts energy innovation in several ways. First, it is evident that notwithstanding DOE's proclaimed strategic goals, partnerships are not central to DOE's strategy, nor are they systematically planned and executed. Second, funding volatility makes evaluation and planning more difficult, as it is impossible to attribute outcomes to actions in such a noisy environment. DOE project and program managers are subjected to such uncertainty that planning and execution can be extremely difficult. And third, for DOE's active and prospective partners in the private sector, uncertainty makes it difficult to plan, discourages participation, and makes assistance less valuable.

4.5. Project Management for Public–Private Partnerships

Innovation is a difficult enterprise that often crosses the boundaries between organizations, industries, and research fields (Carlsson & Stankiewicz, 1991; Garud & Karnøe, 2003). Innovative projects are notoriously difficult to manage (Cohen & Levinthal, 1989; Saeed, 1998; Ngai, Jin, & Liang, 2008). Respondents to the Survey of Energy Innovation (Section 4.4.3) cited difficulties in the process of innovation more than any other type of challenge. This is even truer if we include commercialization cost as an innovation process rather than a market barrier. Our survey and others have found that firms are turning to partnerships as a means of addressing innovation challenges (Cosner, 2010). Technology partnerships can have cost advantages (Hemphill & Vonortas, 2003) and strategic value (Mowery & Teece, 1996; Wessner, 2002) but raise new management challenges (Stiglitz & Wallsten, 1999; Friend, 2006). In sum, management issues can be as important to the success of innovation as resources, technology, and markets.

Experts interviewed for this research agreed that management practices are important. Real partnerships involve sharing resources, risk, and control over project decisions (or discretion). Such relationships require intensive, ongoing collaboration. Drawing on lessons from the literature and from our own research, we have identified a number of principles that the DOE can use to make partnerships and transactions more effective:

- *Choose partners with similar work cultures.* A match in organizational culture between the parties is important for resolving problems during the project, and for developing agreements ahead of time. Common cultural expectations and approaches to dispute resolution make successful projects more likely.

- *Select partners with overlapping interest and complementary abilities.* Parties' interests will not be the same, but they should be aligned in the specific area of the partnership, not just in a vague or general way, but also in terms the priority that each organization puts on the project. They should have similar goals and place comparable priority on achieving them. If partners have matching abilities, there is a higher risk of conflict as each tries to do the same thing. However, complementary abilities result in division of labor, interdependence, and synergies within the project. Combined with common culture, overlapping interests and complementary abilities lead to higher levels of trust through common commitment and mutual oversight.
- *Use partnerships only for high-priority projects.* Partnerships may not be appropriate for every interaction. One informant reports being steered toward a CRADA – a long and detailed agreement involving intellectual property disclosure and an application for federal funding – when a short-term transaction was all that his firm needed or wanted. Mechanisms that are appropriate for a partnership are probably not appropriate for other activities. Work for Others and User Facility agreements allow the national laboratories to "participate with talent" (as one informant put it) or with resources in supporting the priorities of its clients. Grants and contracts allow DOE to use firms' talents to support its priorities. In fact, a set of projects over time, from informal contacts to small transactions and then onto larger ones, may build the relationship between parties that is necessary to having a successful partnership. Relationships and trust should be seen as both inputs to and outcomes of successful projects – precursors to success as well as indicators of success.
- *Prepare detailed agreements that include planning for contingencies and operational details.* Partnership agreements should cover contingencies and definitions in sufficient detail that the boundaries of shared decision making and the risk of losses are clear to both parties. They should agree on accounting systems, chains of command, information flow, and other systems that are unique, even defining, to individual organizations. If private and federal partners have different motives or expectations for participating in a project, partnership agreements should be extensive. Careful planning and design can reduce the likelihood of encountering unexpected obstacles and enable partners to address those that may arise.
- *Plan jointly before and after a partnership has started.* Technological development projects are complex and require detailed planning. Planning has to be a joint process between partners, including milestones, decision points, evaluation criteria, and contingency plans. Ideally, planning should proceed in parallel with negotiating the partnership agreement, as these two aspects are interrelated. The planning should explore multiple paths that could cause a project to fail, and should identify options for abandoning the project or dissolving the partnership if necessary. Understanding failure modes increases the chances of avoiding them, as well as the potential for learning from technological setbacks. Moreover, an amicable dissolution of a partnership leaves open the possibility of future collaboration. In our interviews, subjects recommended spending 10% to 20% of total project time on planning.
- *Focus in the planning process on relationships and scenarios, not on operational details.* Interviewees noted a disconnect between the amount of time it took to begin work within a

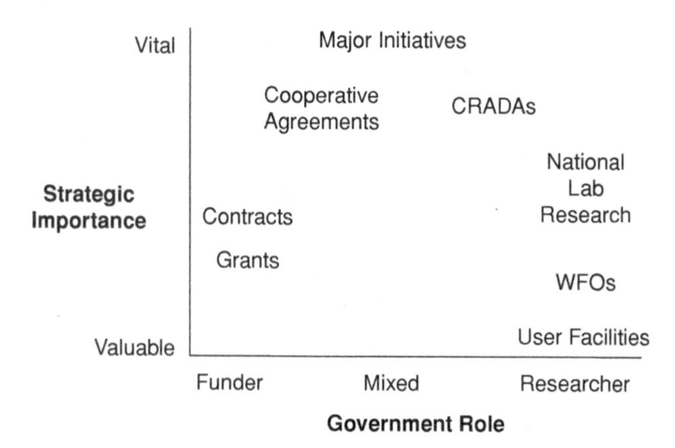

Figure 4.9. Typology of energy RD&D partnership and transaction activities by priority level and type of government involvement.

partnership and the quality of the project plan (often a long planning process did not result in a good plan). In their experience, time spent at the beginning of projects usually included too many financial and intellectual property details and not enough contingency planning and relationship building. An excessive focus on accountability leads to burdensome procedures, accounting systems, and documentation requirements

- *Build trust and relationships between organizations.* Even when partnerships are guided by detailed agreements, trust is essential for success (Teece, 1992). In complex, uncertain technology development work, difficulties are likely to arise and no agreement can cover all possibilities. One group of informants contrasted two recent CRADAs conducted at a national lab with private firms. One was primarily assembled by management: researchers were assigned to the project but were not always involved in project planning. The collaboration lasted only a year and accomplished nothing. A more successful partnership was driven by contacts between researchers. Other collaborations and deals followed. If the balance between relationships and tasks tips too much toward tasks, neither gets accomplished. If it tips too much toward relationships, technical failures may occur but there is more room for learning.
- *Distinguish between technological failure and project failures.* Innovation is an uncertain enterprise and not every technological idea will work. If there is too much pressure to succeed, managers will not take risks and will communicate only their successes. If objectives are too timid, a project will have little benefit even if all objectives are met. An aggressive set of projects can produce valuable knowledge even if many of them fail to produce tangible products (Frosch, 1996; Euchner, 2009). Good planning makes learning more likely. To make use of knowledge, it is important to have a culture where "learning from failure" describes not just individual actors avoiding the repetition of mistakes but also the dissemination of negative results to a wider group.

These observations lead to a typology of activities, as shown in Figure 4.9. Government roles range from funding source to principal researcher, with various degrees of partnerships in between. And some projects will be more directly related to national energy strategy goals than others. These two variables generate a range of appropriate mechanisms for partnerships. The most important work calls for major initiatives, likely to require many partners. For sufficiently important work, it is worth engaging in partnerships through CRADAs or cooperative agreements. Other work may be important, but may be better accomplished through less complicated support or technology transfer mechanisms.

4.6. Strategic Planning in Both Public and Private Sectors, and in Partnerships

To capture the maximum value from investments in energy technology innovation, they must be guided by sound strategies. Elsewhere in this report we discuss strategies for allocating effort (Chapter 2), designing innovative institutions (Chapter 3), and engaging in international cooperation (Chapter 5). In the realm of public–private partnerships, problems are likely to be experienced at the project level as failed projects, timid projects, and technology that does not progress into use. Because successful partnerships are difficult to achieve, planners should treat partnerships as elements of strategy. Doing this effectively requires flexible mechanisms and good information flow.

4.6.1. Current Process for Making Decisions about Energy Innovation Partnerships at DOE

Strategic decisions on funding for energy RD&D within DOE, for each broad technology area, occur in the context of interactions between budget and policy planners and senior program managers. Strategy reflects administration priorities, and also perceived opportunities, hurdles, and each energy technology's potential to increase security of supply, reduce cost and environmental impacts, and increase U.S. global competitiveness.

Program offices, the national laboratories, and technology experts participate in forming technology roadmaps and multiyear plans (e.g., NREL, 2007), which are influenced by, but also inform, department-wide priorities. The multiyear plans help to determine the technological potential of each energy source, which is then used by policymakers at the DOE headquarters level, at OMB, and at the White House to formulate budget requests. Budget documents and program plans offer visible strategic guidance. Though they appear to embody a great deal of technological information, these documents focus primarily on allocating funds among programs and rarely discuss rationales or strategies for partnering with private firms in different technology areas.

DOE's mission is to "ensure America's security and prosperity by addressing its energy, environmental, and nuclear challenges through transformative science and technology solutions." The agency's stated goal is to "catalyze the timely, material, and efficient transformation of the nation's energy system and secure U.S. leadership in clean energy technologies" (DOE, 2011). With its R&D support and technology transfer mechanisms (Section 4.3.1), the DOE can serve both researchers and research funders, and provide unique services to each. The agency acts as a clearinghouse for information on energy, and has grants and load guarantees to support deployment. DOE tries to target its support and services to areas that the private sector cannot sufficiently support on its own.

From the agency's strategic guidance documents and subject to budget constraints and rules, program offices generate funding opportunity announcements. Firms and other potential partners respond to the announcements with proposals and applications, or they submit unsolicited proposals within the published areas of interest.

Proposals are evaluated through peer review. Partnerships are considered inherently valuable, so awards favor proposals with groups collaborating. Based on the data in Section 4.3, 88% of DOE assistance in energy RD&D goes to shared-cost projects (mostly cooperative agreements) with industry. The national laboratories are assigned some work directly, but they also compete for work among themselves, and sometimes they are secondary assets of firms making proposals. In addition, the national laboratories make decisions about technology transfer projects. They work to attract projects that will further strategic goals, but must also rely on proposals from partners.

The current system has strategic and project-level weaknesses that should be addressed department-wide, although as noted some pockets at DOE are implementing some of these recommendations to different extents:

- *Strategic plans do not offer guidance about forming partnerships.* The main strategic guidance on partnerships from DOE includes DOE-specific strategic plans (DOE, 2006, 2011), budget documents (DOE, 2010), multiyear program plans (EERE, 2008), and internal directives. All of these sources state that partnerships are important, but none provide guidance on how to develop them. In our interviews, respondents from government and the private sector were unable to identify documents that explained when or why DOE would use one mechanism (i.e., a grant, a cooperative agreement, etc.) over another. All of these documents contain great detail about technology and goals in terms of volume or performance, but they offer much less information about appropriate types of mechanisms or what kinds of partnerships have worked in the past. Project management occurs in isolation from strategic decisions.
- *Partnerships are not sufficiently considered in budget and strategic decisions.* Budgets and plans are largely developed in isolation from project management, although the types of mechanisms that will be used in a program, as well as expected cost-sharing contributions, affect program cost. And partners' capabilities affect what capabilities need to be developed by DOE.
- *Partnerships are not treated as strategic relationships.* The selection of cooperative agreements and CRADAs relies on a fairly passive announcement–application–review process.

Private firms might seek out and court partners that meet their strategic interests (Hagedoorn, 1993; Hemphill & Vonortas, 2003), but the government is generally prohibited from such practices by competitive bidding rules. The relationship between partners is important for project success (Teece, 1992), but factors such as cultural fit and complementary capabilities do not factor into the decision process until after the projects are selected. The best relationships described by interviewees developed through ad hoc contacts, not by design.

- *Cooperation mechanisms are inflexible.* Unlike private sector technology partnerships, the government cannot promise deliverables and has limited ability to take on risk, so some partners have been disappointed with the results. Agreements are limited in time and scope, thereby preventing the exploration of new avenues for innovation if any are uncovered during the project.
- *Administrative requirements are not streamlined.* Complex administrative requirements affect all stages of the budget and management decision process. This is more relevant at the project level than at the strategic level, but nevertheless requires department-wide action. Informants with private-sector experience describe the applications, agreements, record keeping, and reporting requirements as both more difficult than, and inconsistent with, standard business practices. Detailed planning was cited as an important factor in project success, but this is not synonymous with adding more bureaucracy. For example, negotiators of agreements spend more time on accounting systems than on technical and partnership contingencies.
- *Information collected does not support learning.* As noted in Section 4.3.1, little information is available the technology, development stages, processes, and outcomes associated with them. Current reports are intended for accountability and not for learning, although interviewees questioned the degree to which they even served this purpose.
- *Political barriers can prevent DOE from supporting the best projects.* Finally, informants perceived systemic political barriers to implementing a strategy. Federal entities are prohibited from competing with private enterprise, yet, by design, partnerships with the government provide value to private firms. In some cases, interviews pointed to DOE's inability to provide support for valuable projects for this reason. The review process and availability of services are meant to ensure fairness, but not all parties accept the outcomes. When there are successful partnerships that lead to commercial value, nonparticipant (and sometimes competing) firms have complained that DOE has provided unfair support to its partners in the project. In another example, informants report having experienced pressure from Congressional offices to avoid helping a competitor of their constituents.

Paradoxically, the administrative requirements that govern partnerships are too rigid, but at the same time they lack direction and focus. Opportunities exist to improve the pace of technological innovation through strategic and management improvements. The expanded federal role in funding energy technology innovation under the ARRA (ARRA, 2009) should be matched by a process that supports better management of that enterprise. In Section 4.6.2, we propose changes to accomplish this goal.

4.6.2. Improving DOE Decision-Making Process on Partnerships

Scientists, managers, policymakers at DOE and in the national laboratories, and their partners have done an admirable job promoting energy technology innovation, bringing innovations to market, and developing the innovation system. They can point to many specific successes (DOE, 2007; FLC, 2007a) and have influenced energy technology trajectories (DOE & EPRI, 1997; Norberg-Bohm, 2000; EIA, 2001). However, our interviews show that impediments remain to implementing the kind of energy innovation enterprise required to solve today's major energy-related challenges. Improving the strategic process will require changes to enhance the flow of information, even if less detail is transmitted to the top. Improving the success rate of individual projects will require system-wide changes to improve management.

Knowledge about technology is widely distributed and used for planning, but information about strategy and partnerships is localized at the top and bottom of the management chain. Everyone may read the DOE strategy but it contains little information that can help program offices select or manage projects. There are some interactions between officials who design multiyear program plans and those who make budget decisions, but at the project level managers receive input primarily from above. Project selection is separated from both strategic decisions and project design, and projects are designed only after they are selected.

Strategic decisions on budgets, program plans, and partnerships should be made together in an integrated manner. In an improved process, project design is considered as part of the selection process, not as an afterthought. Project managers receive useful strategic directions, and their experience is transmitted to strategic decision makers to inform policymaking.

Budgets, program plans, and funding opportunity announcements can also be improved in ways that will ensure a closer match between proposals and selected projects on the one hand and the nation's strategic interests on the other. We recommend these principles to improve information flow and address some of the weaknesses in the current process:

- *Articulate how DOE's partnership roles fit with its mission and goals.* DOE acts as a researcher, research funder, technical expert, provider of unique research services, clearinghouse for information, and financier for deployment. For each role, there should be a positive statement of what goals are supposed to be served by that role and how those goals can be reached (Hagedoorn, 1993; Audretsch et al., 2002). Managers will be better able to incorporate strategy into decision making if it is clear what they are meant to achieve through energy innovation support, partnerships, and services. DOE has made some progress in explaining new mechanisms such as Advanced Research Project Agency-Energy (ARPA-E) and Energy Innovation Hubs, but even the metaphors used to explain these initiatives are geared toward Congress rather than managers.

- *Set forth principles for when and how to use different mechanisms.* Government support and services are designed to correct market failures and advance strategic interests, but these goals are only vaguely referred to in strategic documents. Managers can be more deliberate in the mechanisms they choose if those mechanisms match the problems they are intended to address (Hagedoorn, Link, & Vonortas, 2000; Fri, 2003; Martin, 2003). DOE has many options for stimulating energy innovation: it can make research less expensive for firms, or create an incentive for firms to do research, or conduct the research that firms will not conduct. Managers need guidance to help them decide when creating a partnership is justified, and what they can accomplish through funding and other transactions.
- *Use information about strategy for project design and selection.* Funding opportunity announcements should reflect strategic interests related to the project, market needs, and the expected structure of projects, as well as the technology areas. Most activity supporting work in the private sector and academia will not be true partnerships that involve shared resources, risk, and discretion, but managers should decide deliberately when to undertake true partnerships and when to use other mechanisms.
- *Use information about partnerships in program planning.* Multiyear program plans should take into account what capabilities and resources will be needed from private firms, and how they can be best leveraged, based on information collected about previous efforts. Managers' assessments of these issues are as important as the state of technology in deciding what actions to take. Partnerships are good sources of information about the pace of technology development and about roadblocks.
- *Use information about partnerships in the budget process.* Project funding represents a large share of DOE's budget. The agency's roster of partnerships influences resource requirements, and the availability of matching funds and the capabilities provided by offices, labs, and partners all represent costs to be incurred.
- *Develop strategic relationships where strategic interest justifies doing so.* Relationships between private and government partners develop over time, often starting from informal contacts and building into CRADAs or cooperative agreements. An arms-length approach can prevent favoritism, but if the relationship impacts strategic priorities, it may be necessary to be proactive. Being open and transparent about what a partnership entails can help mitigate dissatisfaction on the part of other parties. Building relationships with all industry players should be a strategic goal so that fewer firms will be left out. Good relationships with industry will help produce successful partnerships and communication channels for technology planning.
- *Create a new mechanism to enable learning.* DOE's current record-keeping system for partnerships is both inadequate for planning and learning purposes and cumbersome in terms of administrative requirements. Any new requirements should occur only in parallel with paperwork reduction. A need exists for information that connects project results to strategic, design, and management decisions. This information should be organized in various levels of detail, from summaries of lessons learned to data that can be used to generate summary statistics and tests that will support policy research. To be useful for learning, recording systems should be separate from reward or renewal systems (Paper & Chang, 2005) so that there is no incentive to inflate scores. In such a system, it would be particularly useful to examine lessons learned from projects where technical goals were not met.

- *Simplify administrative requirements and make partnering mechanisms more flexible.* DOE should move toward standard business practices in national laboratories' relationships with private firms by simplifying requirements wherever possible; updating CRADAs, WFO, and other agreements to conform to modern practices; and allowing laboratories to take on risk and make commitments as part of agreements. Because informal contacts and small projects launch relationships that lead to successful partnerships, DOE strategy should explicitly include funding to support activities other than full partnerships and the flexibility for researchers to make contact with industry on an informal and smaller-project level.
- *Allow projects to fail.* If DOE takes on only the most promising projects, it will be doing work that industry could do itself (Martin, 2003). A culture of risk and experimentation is required to find technologies that will transform the energy system (Fri, 2003). Programs should take on a portfolio of projects knowing that some will not work as desired. If a technology does not work, research can still produce new knowledge (Frosch, 1996; Euchner, 2009). If the project itself goes poorly, there is an opportunity to learn management lessons. Managers should not be punished in either case.
- *Allow projects to succeed.* In contrast, stakeholders have to accept that some firms will benefit from successful projects. To alleviate any sense of unfair support, part of the strategic priority in each technology area should be to engage in problems useful to multiple parties, and to share the results widely. DOE should try to engage industry as a whole in developing roadmaps and program plans in the traditions of SEMATECH, Gas Research Institute, and Electric Power Research Institute (EPRI) (see Chapter 3).

Improving the flow of information through these strategic adjustments will align funding and technology transfer opportunities with strategy, as they would result from an interactive communication. Projects would be selected based on (1) technology area, (2) identified needs within a roadmap, (3) overlapping interest within those needs, and (4) desired complementary capabilities. Strategic planning would incorporate a better picture of innovation that considers how projects work in addition to the state of technology. Strategy that supports better management will produce greater satisfaction on the part of partners and greater value for the investments made.

4.7. Learning and Adaptation

In its overall approach to energy technology innovation, DOE should foster a management culture that embraces experimentation and learning, just as it fosters those attitudes toward solving technical problems. Considering partnerships as elements of strategy, engaging in real partnerships only when they are worth the effort, expending the effort in planning needed for project success, and improving the information systems for projects are important steps in leading a successful innovation enterprise. Given that DOE's collaboration with the private sector through grants and cooperative agreements (overall data on R&D consortia were not available) amounted to $4.3 billion for energy RD&D and $1.6 billion for Science, and that DOE made up about one fifth of U.S. investments in energy RD&D including only DOE investments and those documented by the NSF BRDIS, this is an important area.

Innovative organizations are learning organizations (Senge, 1990). DOE is already adept at discovery through research, but learning also involves changing practices in response to information. An organization's ability to learn is, in part, a cultural orientation, and requires occasional renewal to maintain commitment and vision (Calantone, Cavusgil, & Zhao, 2002). Learning organizations must be able to adjust their strategy in response to changing conditions or performance problems (Lant & Montgomery, 1987). To improve its use of strategic partnerships, DOE should be able to adjust in response to:

- *External conditions in the energy industry.* Factors such as energy prices, supply security, industry structure, and market conditions are largely outside of DOE's control. They affect research needs and influence the private sector's interest in and ability to conduct energy innovation.
- *Technology development.* What is known, what is possible, and what needs to be done to get new technologies into service are all conditions that change over time. Progress toward DOE's own goals influences the agency's mission. Technology gets closer to market, and new possibilities appear, but new problems emerge and some ideas turn out to be dead ends. DOE should be able to adjust where it engages in the R&D process, the technologies and fields that it supports, and the methods that it uses to engage.
- *Performance of DOE's programs.* A learning organization monitors and adjusts its own practices and does not blame failures on external factors. DOE will have to make adjustments to program performance to promote continual improvement.
- It is important to adjust strategies around lessons learned and changing conditions, but shifting direction with every discovery, budget, or election cycle would be harmful. Gradual, deliberate, and transparent adjustments are appropriate.

Technology roadmapping is an important source of information about technology and industry factors, both external and internal to DOE. Roadmapping as a practice for innovation institutions is discussed in Chapter 3, and is already a practice for developing program plans and national laboratory plans. Following the best practices of roadmapping is important for innovation. Good practices include concentrating on barriers and how to overcome them, estimating the potential of and common value of different steps that might be taken, and avoiding research wish lists (Garcia & Bray, 1997; Fri, 2003; Whalen, 2007).

In the context of learning about management, partnerships, and private sector's needs, the roadmapping process is a good opportunity to build relationships with industry. In this process, industry has the opportunity to help shape DOE's strategic direction, and in so doing, reduce the chance of political problems referred to in Section 4.7.1. Roadmapping in the development of program plans should include the kind of projects and the appropriate mechanisms to be used – that is, it should specify how to work on problems, not just what barriers exist. To improve learning value, the roadmapping process should consider the effects of prior plans and how recent energy innovation projects have changed the state of technology.

Though roadmapping is an excellent tool for information and learning for technology areas, there is a need for a more systemic equivalent, such as that which was initiated by the *Quadrennial Energy Review* proposed by the President's Council of Advisors on Science and Technology (PCAST, 2010). A periodic review and effort to gather and disseminate comparable evaluations across the department should be implemented to assess the impact of previous programs on the nation's energy system. Learning requires information about actions and outcomes with enough detail to make connections between the two. Information on DOE energy RD&D projects across the department does not seem to be collected (let alone distributed) in a comparable manner. As a result, program managers cannot make inferences about the relationships between project structures and outcomes. The agency's Inspector General and Office of Management occasionally conduct reviews of programs and projects, but these are forensic-style reports that are designed to determine whether actions were in accordance with policy or what went awry with a failed project. It is not clear whether findings from these reports are widely used for training, or if they help managers design better programs. DOE has spotlighted some positive examples (DOE, 2007) and inserted those cases into strategic documents (DOE, 2006, 2008, 2011) but without sufficient detail to support informed decisions. Project management is learned apparently only through individual experience.

A more useful learning program would collect information at various levels of detail in forms that would be usable for different consumers. Record keeping should be an integral part of project management from solicitation and planning through completion, so that managers do not have to rely on only their memory at the end. At the highest level of abstraction, basic variables that could be used for summarizing DOE activities and generating statistics should be kept in a database as easy to access as USA Spending (OMB, 2010). Variables such as technology area, stage of development, type of agreement, program and subprogram, outcome measures, and so on are not included in the USA Spending database and do not seem to be collected consistently across DOE (there are exceptions in specific Offices) but would be valuable in measuring DOE activities. Ideally, these data could be linked with financial data required by OMB, perhaps through Federal Award ID numbers. Using these data to infer trends and problems should be a part of policymaking and oversight.

Managers at the program and project level need greater detail about what has and has not worked in past DOE energy RD&D projects. To enable managers to learn from each other's experience, it should be standard practice to report on and catalog lessons learned or project narratives. These reports should include relationships with partners, problems that arise and their solutions, contributions and expectations from each party, scientific questions posed and answered, and project outcomes. Reports should be archived in a searchable database. And DOE should encourage managers to use the information to improve their own knowledge – particularly to learn from peers outside their own area.

Personnel in the Office of Management or the Office of Policy would be able to use the reports to research best practices in various subject areas. To be useful, these reports should not be used against managers or partners in performance evaluations. Emphasis should be placed on writing useful reports, especially analyses of "failed" projects.

Learning organizations use internal and external information to learn from the past and adjust to changing conditions. Staff members are free to discuss problems without consequences, and failures are viewed as opportunities for improvement. Learning and adaptation take place constantly. Understanding of what constitutes best practice changes over time as technology develops and new problems constantly emerge. A learning organization has to re-commit itself to continuous improvement periodically, and even has to monitor and adapt how it learns, and how it learns how to learn.

4.8. Recommendations

Consistent with the findings of other studies (Grubb, Köhler, & Anderson, 2002; OECD & IEA, 2003; Mowery, Nelson, & Martin 2010; PCAST 2010), we conclude that the most effective single factor for driving innovation in the private sector would be a price on, or a market for, greenhouse gas emissions. Our interviews and research support the view that private firms respond to price signals by innovating. This conclusion reinforces our finding in Chapter 2 that without a carbon price or other demand-side policies, CO_2 emissions are unlikely to decrease to the levels that will be needed to stabilize atmospheric concentrations at a level that will avert drastic impacts from climate change. A carbon tax or a limit on CO_2 emissions implemented through an economy-wide cap and trade program, or as a second-best policy such as a clean energy standard for the power sector, could have a significant impact on the extent and intensity of private sector innovation in energy, and therefore on the cost of mitigating CO_2 emissions.

As a corollary, the U.S. government must support innovation to enable the market to work with acceptable costs, and to speed up the transition to a new energy system. Supporting the private sector and working in public–private partnerships will contribute to a clean energy transition. To do so more effectively, DOE should:

- *Include partnerships in strategic decisions.* Set forth how roles, support mechanisms, and partnerships fit in with national priorities. If a role does not fit into the missions and goals of national priorities, it should be moved outside DOE. If a mechanism does not serve DOE's needs it should be changed. If a capability is missing it should be developed. Provide guidance on when to use partnerships and what kind of support suits what kind of innovation.
- *Become a learning organization.* Collect data and information on practices, problems, solutions, and outcomes, and use that information to set policies and educate managers.

Create an atmosphere in which people feel free to discuss problems, and that prioritizes adaptation and learning.

- *Build a strategic relationship with industry*. Use roadmapping and other meetings to collect information on what the private sector needs and build support for DOE efforts. Because the private sector is diverse, provide services and promote innovation through initiatives tailored for each particular industrial sector. Consider a member/advisor/ownership model for interaction with industry, like the Semiconductors Research Corporation member board. Understand that markets and short-term decisions are important for firms and adapt interventions to account for their decisions. In addition, DOE should establish an Energy Innovation Advisory Board (EIAB) with representation from the private sector, academia, the national laboratories, and other key stakeholders, to advise on how best to accelerate private sector energy technology innovation. This advisory board would supplement the existing Secretary of Energy Advisory Board by providing a specific focus on policies designed to accelerate the pace of improvement in energy technologies in collaboration with the private sector. It would also complement and inform the *Quadrennial Technology Review*, which was conducted for the first time in 2011. The board's mission should include consideration not only of DOE investments in energy technologies, but also of the broader set of U.S. government policies that affect the pace of energy technology innovation.
- *Use strategy in partnership decisions*. Match the type of support to the needs and priority of the project. Write Funding Opportunity announcements to attract the kinds of partners that suit strategic interests. Improve planning and relationship building to increase the chance of project success, and work to learn from problems. Engage in partnerships only when it serves a strategic priority, and recognize the value of other forms of support.
- *Expand information about the private sector*. A follow-up to the *Survey of Energy Innovation* will provide valuable data for serving private sector needs. The proposed *Quadrennial Energy Review* can provide valuable information about the needs of the energy innovation system. Contacts with industry can provide information about technological barriers and opportunities as well as the needs of industry.
- *Be strategic about information*. Whatever data are collected from all these sources, as well as internal data, must be kept in a way that is accessible to internal managers and external researchers in order to be of the greatest use.

Together, these recommendations can help build a U.S. government approach to working with the private sector on energy innovation that meets the CASCADES criteria for an effective energy innovation strategy outlined in Chapter 1. Such a strategy must be *comprehensive*, focusing on multiple stages of innovation and multiple institutions and stakeholders. In particular, it must include both investments to create technology push and deployment policies to create demand pull. It must be *adaptive*, learning from experience and making adjustments accordingly. It must be *sustainable*, for industry will not respond well to policies that flit from one priority to the next as officials come and go. Such policies are more likely to gain private-sector support, and to be politically sustainable, if they are *cost-effective*, gaining as much progress as practicable per dollar invested. The need for these efforts to be *agile*, shifting to seize new opportunities as they arise, was not a central theme of this chapter, though it is important and was addressed in Chapter 3. Just as private firms diversify their investments, these efforts should be *diversified*, covering a broad range of technological possibilities. They must be *equitable*, providing comparable support to different technologies and approaches, rather than investing more in one area simply because one industry group has more influence on government policymakers than another. And it is clear that these investments must be *strategic*, focusing on working with the private sector to achieve key goals over the long haul.

Ultimately, the most important task is to build a process for adapting strategy to changing conditions and lessons learned. No optimal strategy can be followed forever. Just as important, strategy has to adjust smoothly and transparently rather suddenly with every election or budget cycle. Learning occurs if questions being asked about technology change in any direction. Leaders of the energy technology innovation system have to monitor the pace of learning and adapt strategy to improve it.

References

Anadon, L.D., Gallagher, K.S., Bunn, M., & Jones, C.A. (2009). *Tackling U.S. Energy Challenges and Opportunities: Preliminary Policy recommendations for Enhancing Energy Innovation in the United States*. Cambridge, Mass.: Energy Technology Innovation Policy Research Group, Belfer Center for Science and International Affairs, John F. Kennedy School of Government, Harvard University.

Anadon, L.D., Bunn, M., Chan, G., Chan, M., Jones, C., Kempener, R. Lee, A, Logar, N., & Narayanamurti, V. (2011). *Transforming U.S. Energy Innovation*. Cambridge, Mass.: Report for Energy Technology Innovation Policy Research Group, Belfer Center for Science and International Affairs, John F. Kennedy School of Government, Harvard University.

ARRA. (2009, February). American Recovery and Reinvestment Act of 2009, P.L. 111–5, 111th Congress, session 1.

Arrow, K.J. (1962). "Economic welfare and allocation of resources." In *The Rate and Direction of Inventive Activity: Economic and Social Factors*, edited by R.R. Nelson, pp. 609–26. Cambridge, Mass.: National Bureau of Economic Research, Inc.

Audretsch, D.B., Bozeman, B., Combs, K.L., Feldman, M., Link, A.N., Siegel, D.S., Stephan, P., Tassey, G., & Wessner, C.W. (2002). "The economics of science and technology." *Journal of Technology Transfer*, 27:155–203.

Battelle. (2010). "2011 global R&D funding forecast." *R&D Magazine*. Elk Grove Villiage, IL. Retrieved from http://www.battelle.org/aboutus/rd/2011.pdf (Accessed March 15, 2011).

Block, F., & Keller, M.R. (2008). *Where Do Innovations Come from? Transformations in the U.S. National Innovation System, 1970–2006*. Washington, D.C.: The Information Technology & Innovation Foundation.

Bloomberg New Energy Finance, (2013). "Global trends in renewable energy investment 2013." Retrieved from http://www.unep.org/pdf/GTR-UNEP-FS-BNEF2.pdf (Accessed February 12, 2014).

Brick, J.M., Morganstein, D., & Valliant, R. (2000). *Analysis of Complex Sample Data Using Replication*. Report. Westat. Retrieved from http://www.westat.com/westat/pdf/wesvar/acs-replication.pdf (Accessed February 12, 2014).

Brown, M.A. (2001). "Market failures and barriers as a basis for clean energy policies." *Energy Policy*, 29:1197–1207.

Calantone, R.J., Cavusgil, S.T., & Zhao, Y. (2002). "Learning orientation, firm innovation capability, and firm performance." *Industrial Marketing Management*, 31(6):515–24.

Carlsson, B., & Stankiewicz, R. (1991). "On the nature, function, and composition of technological systems." *Journal of Evolutionary Economics*, 1(2):93–118.

CCTP. (2006). *U.S. Climate Change Technology Program Strategic Plan*. Washington, D.C.: Climate Change Technology Program.

Census. (2009). "Business R&D and innovation survey." U.S. Census Bureau. Retrieved from bhs.econ.census.gov/BHS/BRDIS/index.html (Accessed December 18, 2009).

Cohen, W.M., & Levinthal, D.A. (1989). "Innovation and learning: The two faces of R & D." *Economic Journal*, 99(397):569–96.

Cosner, R.R. (2010). "Industrial Research Institute's R&D trends forecast for 2010." *Research Technology Management*, 53(1):14–22.

DOE. (2005). *Audit Report: Selected Energy Efficiency and Renewable Energy Projects*. Washington, D.C.: U.S. Department of Energy, Office of Inspector General, Office of Audit Services. May DOE/IG-0689. Retrieved from http://energy.gov/sites/prod/files/igprod/documents/CalendarYear2005/ig-0689.pdf (Accessed February 12, 2014).

DOE. (2006). *U.S. Department of Energy Strategic Plan*. Washington, D.C.: Department of Energy.

DOE. (2007). *Report on Technology Transfer and Related Technology Partnering Activities at the National Laboratories and Other Facilities, Fiscal Year 2006*. Washington, D.C.: Department of Energy.

DOE. (2008). *Annual Procurement and Financial Assistance Report, FY 2008*. Washington, D.C.: Department of Energy.

DOE. (2009a). *Agency Financial Report*. Washington, D.C.: Department of Energy.

DOE. (2009b). *Annual Procurement and Financial Assistance Report, FY2009*. Washington, D.C.: Department of Energy.

DOE. (2009c). "R&D support." Department of Energy. Retrieved from www.doe.gov/r&dsupport.htm (Accessed August 19, 2009).

DOE. (2009d). "SBIR & STTR." Department of Energy. Retrieved from sbir.er.doe.gov/sbir (Accessed August 19, 2009).

DOE. (2010). FY 2011 Congressional Budget Request: Budget Highlights. Washington, D.C.: Department of Energy.

DOE. (2011, September). Department of Energy Quadrennial Technology Review. Washington, D.C.: Department of Energy. Retrieved from http://energy.gov/sites/prod/files/QTR_report.pdf (Accessed February 12, 2014).

DOE & EPRI. (1997). *Renewable Energy Technology Characterizations*. Washington, D.C.: Department of Energy and Electric Power Research Institute.

Dooley, J.J. (2010). *The Rise and Decline of U.S. Private Sector Investments in Energy R&D Since the Arab Oil Embargo of 1973*. Springfield, Va.: NTIS.

EERE. (2008). *Building Technologies Program Planned Activities for 2008–2012*. Washington, D.C.: Office of Energy Efficiency and Renewable Energy.

EIA. (2001). *Renewable Energy 2000: Issues and Trends*. Washington, D.C.: Energy Information Administration.

EIA. (2009). "Energy & finance." Energy Information Administration. Retrieved from eia.doe.gov/emeu/finance/index.html (Accessed July 13, 2009).

EIA. (2010). *Annual Energy Review*, 2009. Washington, D.C.: Energy Information Administration.

EIA. (2012, September). *Annual Energy Review 2011 (DOE/EIA-0384(2011)* Washington D.C.: Government Printing Office.

Energy Innovation Tracker. (2013). "Energy innovation tracker: About us." Retrieved from http://energyinnovation.us/about/ (Accessed August 21, 2013).

EOP. (2009). *A Strategy for American Innovation: Driving Towards Sustainable Growth and Quality Jobs*. Washington, D.C.: Executive Office of the President.

Euchner, J.A. (2009). "Innovation's family tree." *Research Technology Management*, 52 (6):9.

Farris, W. (2010). "FY09 Department of Energy technology transfer data." [Personal communication, April 10, 2010]

FLC. (2007a). *Federal Laboratories & State and Local Governments: Partners for Technology Transfer Success*. Cherry Hill, N.J.: Federal Laboratory Consortium.

FLC. (2007b). *Technology Transfer Mechanisms Used by Federal Agencies: A Quick Reference Guide*. Cherry Hill, N.J.: Federal Laboratory Consortium.

Fri, R.W. (2003). "The role of knowledge: Technological innovation in the energy system." *The Energy Journal*, 24(4):51–74.

Friend, J.K. (2006). "Partnership meets politics: Managing within the maze." *International Journal of Public Sector Management*, 19(3):261–77.

Frosch, R.A. (1996). "The customer for R&D is always wrong!" *Technology Management*, 39(1):22–7.

Gallagher, K.S., & Anadon, L.D. (2010). *DOE Budget Authority for Energy Research, Development, and Demonstration Database*. Cambridge, Mass.: Energy Technology Innovation Policy Report.

Gallagher, K.S., Holdren, J.P., & Sagar, A.D. (2006). "Energy-technology innovation." *Annual Review of Environment and Resources*, 31:193–237.

GAO. (2009, June). "Clearer priorities and greater use of innovative approaches could increase the effectiveness of technology transfer at Department of Energy laboratories." Washington, D.C. Retrieved from http://www.gao.gov/products/GAO-09-548 (Accessed February 12, 2014).

Garcia, M.L., & Bray, O.H. (1997). *Fundamentals of Technology Roadmapping*. Albuquerque, N.M.: Sandia National Laboratories.

Garud, R., & Karnøe, P. (2003). "Bricolage vs. breakthrough: Distributed and embedded agency in technological entrepreneurship." *Research Policy*, 32(2):277–301.

Grubb, M., Köhler, J., & Anderson, D. (2002). "Induced technical change in energy and environmental modeling: Analytic approaches and policy implications." *Annual Review of Energy and the Environment*, 27:271–308.

Hagedoorn, J. (1993). "Understanding the rationale of strategic technology partnering: Interorganizational modes of cooperation and sectoral differences." *Strategic Management Journal*, 14(5):371–85.

Hagedoorn, J., Link, A.N., & Vonortas, N.S. (2000). "Research partnerships." *Research Policy*, 29(4):567–87.

Hemphill, T.A., & Vonortas, N.S. (2003). "Strategic research partnerships: A managerial perspective." *Technology Analysis & Strategic Management*, 15(2):255–71.

Jaffe, A.B. (1998). "The importance of "spillovers" in the policy mission of the Advanced Technology Program." *Journal of Technology Transfer*, 23(2):11–19.

Jaffe, A.B., Newell, R.G., & Stavins, R.N. (2005). "A tale of two market failures: Technology and environmental policy." *Ecological Economics*, 54(2–3):164–74.

Kammen, D.M., & Pacca, S. (2004). "Assessing the costs of electricity." *Annual Review of Environment and Resources*, 29:301–44.

Lant, T.K., & Montgomery, D.B. (1987). "Learning from strategic success and failure." *Journal of Business Research*, 15(6):503–17.

Margolis, R.M., & Kammen, D.M. (1999). "Evidence of under-investment in energy R&D in the United States and the impact of federal policy." *Energy Policy*, 27:575–84.

Martin, S. (2003). "The evaluation of strategic research partnerships." *Technology Analysis & Strategic Management*, 15(2):159–76.

Merrill, S.A. (1984). "The politics of micropolicy: Innovation and industrial policy in the United States." *Policy Studies Review*, 3(3–4):445–52.

Mowery, D.C., Nelson, R.R., & Martin, B.R. (2010). "Technology policy and global warming: Why new policy models are needed (or why putting new wine in old bottles won't work)." *Research Policy*, 39:1011–23.

Mowery, D.C., & Teece, D.J. (1996). "Strategic alliances and industrial research." In *Engines of Innovation: U.S. Industrial Research at the End of an Era*, edited by R.S. Rosenbloom, pp. 111–30. Cambridge, Mass.: Harvard Business School Press.

Narayanamurti, V., Anadon, L.D., Breetz, H., Bunn, M., Lee, H., & Mielke, E. (2011). *Transforming the Energy Economy: Options for Accelerating the Commercialization of Advanced Energy Technologies*. Cambridge, Mass.: Belfer Center for Science and International Affairs, John F. Kennedy School of Government, Harvard University.

Narayanamurti, V., Anadon, L.D., & Sagar, A.D. (2009). "Transforming energy innovation." *Issues in Science and Technology*, Fall:57–64.

Narayanamurti, V., Odumosu, T., & Vinsel, L. (2013). "RIP: The basic/applied research dichotomy." *Issues in Science and Technology*, XXIX.2 (Winter) Retrieved from http://www.issues.org/29.2/Venkatesh.html (Accessed February 12, 2014).

Nemet, G.F., & Kammen, D.M. (2007). "U.S. energy research and development: Declining investment, increasing need, and the feasibility of expansion." *Energy Policy*, 35(1):746–55.

Ngai, E.W., Jin, C., & Liang, T. (2008). "A qualitative study of interorganizational knowledge management in complex products and systems development." *R&D Management*, 38(4):421–40.

Norberg-Bohm, V. (2000). "Creating incentives for environmentally enhancing technological change: Lessons from 30 years of U.S. energy technology policy." *Technological Forecasting and Social Change*, 65:125–48.

NREL. (2007). *Wind Energy Multiyear Program Plan for 2007–2012*. Washington, D.C.: National Renewable Energy Laboratory.

NSF. (2009). *Research and Development in Industry, 2005*. Arlington, Va: National Science Foundation Division of Science Resource Statistics.

NSF. (2010a). *National Patterns of R&D Resources: 2008 Data Update*. Arlington, Va.: National Science Foundation Division of Science Resource Statistics.

NSF. (2010b). *Science and Engineering Indicators, 2010*. Arlington, Va.: National Science Foundation.

NSF. (2012). *Business R&D Performed in the United States Cost $291 Billion in 2008 and $282 Billion in 2009*. NSF 12–309. March. Arlington, Va.: National Science Foundation.

NSF. (2013). "Research and development national trends and international comparisons." In *National Science and Engineering Indicators 2012*. Arlington, Va.: National Science Foundation Division of Science Resource Statistics.

NSF. (2013). *National Patterns of R&D Resources: 2010–11 Data Update (NSF13–318)* Arlington, Va.: National Science Foundation Division of Science Resource Statistics.

OECD & IEA. (2003). *Technology Innovation, Development and Diffusion*. Paris: Organisation for Economic Co-operation and Development & International Energy Agency.

OMB. (2010). USA Spending. Web site, Office of Management and Budget. Retrieved from www.usaspending.gov (Accessed June 1, 2010).

Paper, D., & Chang, R. (2005). "The state of business process reengineering: A search for success factors." *Total Quality Management & Business Excellence*, 16(1):121–33.

PCAST. (2010). *Report to the President on Accelerating the Pace of Change in Energy Technologies through an Integrated Federal Energy Policy*. Washington, D.C.: President's Committee of Advisors on Science and Technology.

Saeed, K. (1998). "Maintaining professional competence in innovation organizations." *Human Systems Management*, 17(1):69–87.
Schumpeter, J.A. (1934). *The Theory of Economic Development*. Cambridge, Mass.: Harvard University Press.
Senge, P.M. (1990). *The Fifth Discipline: The Art and Practice of the Learning Organization*. New York: Doubleday/Currency.
Solow, R.M. (1957). "Technical change and the aggregate production function." *Review of Economics and Statistics*, 39(3):312–20.
Stiglitz, J.E., & Wallsten, S.J. (1999). "Public-private technology partnerships: Promises and pitfalls." *American Behavioral Scientist*, 43(1):52–73.
Stoneman, P., & Diederen, P. (1994). "Technology diffusion and public policy." *Economic Journal*, 104(425):918–30.
Teece, D.J. (1992). "Competition, cooperation, and innovation: Organizational arrangements for regimes of rapid technological progress." *Journal of Economic Behavior and Organization*, 18(1):1–25.
Transparency Act. (2006, September). Federal Funding Accountability and Transparency Act of 2006, P.L. 109–282, 109th Congress, session 2.
Unruh, G.C. (2000). "Understanding carbon lock-in." *Energy Policy*, 28(12):817–30.
Wessner, C.W. (2002). *Government-Industry Partnerships for the Development of New Technologies*. Washington, D.C.: National Academies Press.
Whalen, P.J. (2007). "Strategic and technology planning on a roadmapping foundation." *Research Technology Management*, 50(3):40–51.
Wiesenthal, T., Leduc, G., Schwarz, H., & Haegeman, K. (2009). *R&D Investment in the Priority Technologies of the European Strategic Energy Technology Plan*. Seville, Spain: Joint Research Centre of the European Commission.
Wolfe, R.M. (2010). *U.S. Businesses Report 2008 Worldwide R&D Expense of $330 Billion: Findings from New NSF Survey*. Arlington, Va.: National Science Foundation Division of Science Resource Statistics.

5

Maximizing the Benefit from International Cooperation in Energy Innovation

RUDD KEMPENER, MATTHEW BUNN, AND LAURA DIAZ ANADON

5.1. A World of Competition and Cooperation

Energy technology innovation is not a U.S. phenomenon; it is a global phenomenon. All over the world, research institutions, companies, and governments alike are intensifying their investments in energy research, development, demonstration, and deployment (ERD3).

In 1999, a report by the President's Committee of Advisors on Science and Technology (PCAST) called for more than doubling U.S. federal spending on international ERD3 cooperation and for establishing new institutions to develop a strategy and to oversee the effort (PCAST, 1999). It recommended a wide range of international ERD3 cooperation activities over a large number of different energy technologies.

The international energy landscape has been transformed since then, in particular by the rise of emerging economies such as China and India both as energy markets and as energy innovators. Overall, these changes have only made the case for such cooperation stronger. Indeed, the process of innovation has globalized, with ideas, resources, and manufacturing coming from different countries throughout the value chain (Gallagher, Gruebler, Kuhl, Nemet, & Wilson, 2012); there are now few high-tech products that do not include contributions from multiple countries. Countries and firms that attempt to go it alone in developing new energy technologies and bringing them to market are likely to be left behind by those who are sharing in the global flow of ideas and resources.

International ERD3 cooperation therefore is inevitable – the question is not whether to cooperate internationally but how much, with whom, on what, and how. Such international cooperation can be a powerful tool for serving the common interest in energy innovation – drawing ideas and experiences from many sources, pooling resources, taking advantage of the best locations to carry out R&D projects or technology demonstrations, and more. The title of the 1999 PCAST report tells the story: *Powerful Partnerships* (PCAST, 1999). Technology cooperation with institutes and firms in other countries can also provide crucial pathways into the markets in the countries where the cooperation takes place.

Nevertheless, cooperation is always being balanced with competition. Countries around the world have both common and conflicting interests in energy innovation (Levi, Economy, O'Neil, & Segal, 2010). Low-cost clean energy technologies invented anywhere can serve U.S. and global interests through: reduced climate change risks as other countries deploy low-carbon technologies; increased opportunities to develop energy technologies with improved cost and performance without the United States bearing the whole cost of doing so; reduced global dependence on constrained oil resources (potentially leading to lower prices, improved energy security, and less risk of resource conflicts); reduced energy poverty; and potentially a reduced risk of nuclear proliferation. Some argue, for example, that the United States should be happy that Chinese firms are selling solar panels at low prices, offering opportunities for reduced-cost solar deployments in the United States and around the world. But at the same time, countries are seeking to maximize their share of the multi-trillion-dollar energy technology markets of the next few decades, creating a competitive interest. A better, cheaper technology made elsewhere would pose a competitive threat to U.S. firms, and the manufacturing jobs from producing it would go to others, not to U.S. workers; hence the U.S. government and European governments are *not* happy about cheap Chinese solar exports and have been accusing China of unfair trade practices.

Intensified competition in clean energy technologies is here to stay; the United States must compete or it will be left behind. But cooperation can occur in the midst of competition – and indeed, can strengthen the competitive position of the countries that are pooling their resources to make new breakthroughs.

The challenge for the U.S. government in this environment is to balance cooperation and competition to maximize the net benefit for U.S. and global interests. The U.S. government has to face this challenge in a context in which global energy use and energy innovation are rapidly changing; most international ERD3 cooperation is not under government control; and countless programs involving many agencies of the U.S. government, multiple worthy objectives, and scores of countries around the world are already in place. This creates a strategic management challenge of stunning complexity. Ultimately, in an increasingly complex and fast-paced international energy innovation environment, the best the U.S. government can hope to do is to build a supportive international environment for private energy innovation cooperation, let those government officials at the front lines of energy innovation push a variety of initiatives from the bottom up, and push a few key strategic priorities from the top down. In this chapter, we offer recommendations for each of those elements of an overall approach.

First, we describe the changing global landscape of energy production, use, and innovation, focusing on the rise of emerging economies such as China and India as both major energy users and major energy innovators (Section 5.2). These two countries accounted for 61.9% of the growth in global primary energy consumption between 2000 and 2010 (BP, 2011), and at the same time their public expenditure on energy innovation is huge and continues to increase (Gallagher et al., 2011). Next, we outline the changing picture of international ERD3 cooperation, including shifts in energy technology markets and in cooperation among scientists and engineers publishing in the leading energy technology journals (Section 5.3). After that, we describe the myriad cooperation efforts organized by the U.S. government (Section 5.4). This description is followed by a discussion of criteria the U.S. government uses or could use for selecting, executing, and evaluating international ERD3 cooperation (Section 5.5). The chapter ends with a set of recommendations for strengthening the U.S. approach to international ERD3 cooperation and competition, to maximize the opportunities while minimizing the risks (Sections 5.6 and 5.7).

The analysis in this chapter is based on literature reviews, semistructured interviews, and quantitative databases. We used three existing quantitative databases and built two new databases for our analysis of how international ERD3 cooperation between scientists, firms, and governments has evolved over time. These databases include:

- A database we built that focuses on international co-authorship in key journals focused on energy technologies
- A database we built that focuses on U.S. involvement in the International Energy Agency (IEA)'s Implementing Agreements between 1974 and 2009
- An analysis of import and export flows of energy technologies between the United States and other countries around the world between 1989 and 2009, based on a database provided by the United States International Trade Commission (USITC)
- An overview of international interfirm R&D partnerships on energy technologies between 1965 and 2006 based on data from the Cooperative Agreements and Technology Indicators (CATI) database developed by Hagedoorn (2006)
- An analysis of overseas development aid (ODA) for "energy" projects between 2000 and 2009 based on OECD's Creditor Reporting System database (OECD, 2010).[1]

5.1.1. The Critical Importance of the Private Sector

Governments have long played a major role in energy technology innovation, as discussed elsewhere in this book, and they play a major role in shaping international cooperation on such technologies as well. But government-led ERD3 cooperation is only one part of the broader picture of international flows of ERD3 effort. Many of the most essential actions are being undertaken by private players – firms, universities, non-government organizations (NGOs), and individual teams of scientists

[1] A detailed discussion of the features, limitations, and results of these databases can be found in Appendix A5.2 in Anadon et al. (2011).

and engineers. Although there are limited data available, it is clear that private firms' investment in ERD3 efforts that span national boundaries far exceeds U.S. government investments in international ERD3 cooperation.

Private sector cooperation efforts will generally be larger, more dynamic, and more responsive to immediate private-sector needs than government efforts will be. (The complicating factor in making this distinction is the large role played by state-owned firms, particularly in emerging economies such as China and Russia, where they dominate the energy markets and ERD3 efforts.) One of government's most important roles is to enable this bubbling flow of bottom-up cooperation and to avoid standing in the way, limiting the obstacles posed by tariffs, constraints on foreign investment, export controls, visa restraints, and the like. The surest way to stop any technical cooperation is to require that governments must negotiate a legally binding agreement covering every step to be taken before it can begin.[2]

By its nature, such bottom-up coordination is not centrally controlled and tracked, and therefore data on it is difficult to collect and very diverse. Private sector involvement in international ERD3 cooperation can include a large range of activities, such as establishing foreign R&D centers, creating international joint ventures, forming technology alliances, cross-border marketing, establishing foreign subsidiaries, and making foreign direct investments (Beamish & Killing, 1996). Unfortunately, there is little quantitative information available on how patterns of these activities have changed over time[3]; anecdotally, however, it appears that private sector ERD3 cooperation is also on the rise, and increasingly involves partners in emerging economies.

5.2. The Changing Global Landscape of Energy Use and Innovation

The world's energy use is changing rapidly, as are the kinds of investments in ERD3 being made around the world, and the kinds of international collaboration that are underway.

5.2.1. Dramatic Shifts in Energy Production and Use

In recent years there have been amazing changes in both energy use and energy production around the world (IEA, 2012). The bulk of the expansion in energy use is coming from the developing world, particularly major emerging economies such as China and India. At the same time, energy production is shifting as well, with new

[2] For example, Table 3.1 in Chapter 3 shows how government involvement in two of SRC's research programs limits the institution's ability to conduct research on a global scale.

[3] The CATI database, developed by Prof. Hagedoorn, is one of the few databases that provide longitudinal data on international interfirm R&D partnerships. It includes only those partnerships with mutual technology transfer or joint R&D. In collaboration with Prof. Hagedoorn, our quantitative analysis includes the results of this database (Hagedoorn, 2006).

technologies allowing for expanded production of both oil and gas, putting the United States on track to surpass Saudi Arabia as the world's largest producer within a few years. In particular:

- Energy demand in much of the developed world is stable or declining, while it is surging in the developing world. China has replaced the United States as the world's largest energy consumer. China and other non-OECD countries currently account for the majority of the energy use in the world, and all but 4% of the growth in world primary energy demand between 2010 and 2035 is expected to come outside the Organisation for Economic Co-operation and Development (OECD) (Gallagher et al., 2011; IEA, 2012, p. 56). Indeed, almost all projected growth of energy demand over the next quarter of a century will take place in non-OECD countries, making their markets – and the cooperation that may help ensure the U.S. role in them – particularly crucial. While many energy technology vendors used to design their systems for the U.S. market, which was the world's largest, today systems are being designed for China, India, and elsewhere in the developing world.
- Technological improvements have made it possible to find and recover substantial new supplies of oil and gas, from shale gas and tight oil to oil sands, likely turning the United States into a net oil exporter by the 2030s (WEO, 2012). Low-cost natural gas in North America, in particular, has created stiff competition for other energy sources, from coal to nuclear to renewables – and thereby made it more difficult to raise money to develop new energy options.
- The world continues to be overwhelmingly reliant on fossil fuels, which are used to provide more than 80% of global primary energy supply, and under current policies would likely still be providing some three-quarters of a substantially larger global supply in 2035. Remarkably, the use of coal, the dirtiest of the fossil fuels, has grown more than any other energy source over the decade leading up to 2012, meeting nearly half of the total growth in demand (IEA, 2012).
- Global concerns over both carbon emissions and other environmental pollution from energy – particularly the fine particulates from burning coal and diesel fuel, which lead to millions of premature deaths each year – continue to be high, and more and more countries are taking actions to restrain emissions. From energy efficiency to renewables to increased use of natural gas, many steps would reduce both harmful local emissions and greenhouse gas emissions, creating win–win opportunities.
- The use of renewable energy sources is growing at a very rapid rate in percentage terms, but they still provide only a small part of global primary energy demand, and will remain far less than a majority for decades to come (IEA, 2012).
- As more and more countries see clean energy as a critical part of their future, the debate over intellectual property rights and trade rules for energy technologies has intensified.
- The prospects for nuclear energy have dimmed after the Fukushima Daiichi accident in 2011. Japan is struggling to bring its nuclear reactors back on line amid intense public skepticism and protest, while Germany is working to phase nuclear power out entirely. Growth continues in the largest markets (China, India, Russia, and South Korea) after an accident-induced pause. Strengthened safety measures around the world have modestly increased nuclear costs – and estimated capital costs of new nuclear plants continue to grow.

Collectively, these changes mean both good news and bad news for development and deployment of new or improved energy technologies. The continued availability of low-cost coal, oil, and natural gas, coupled with the massive institutional and physical infrastructures in place for these technologies, is likely to mean that fossil fuels will continue to dominate the energy picture and pose stiff competition for other technologies for decades to come. But the global concern over the environmental and security consequences of undue reliance on fossil fuels is likely to drive continued investment in energy innovation. And as seen in Section 5.2.2, the focus of those investments is shifting.

5.2.2. The New Face of Energy Innovation: The Rise of the Emerging Economies

Today, more countries are investing significant resources in energy technology innovation than ever before, and the emerging economies are leading the way. A small group of emerging economies, dominated by China, is now spending more on energy innovation than are all of the developed countries put together. These expanded investments, coming from a more diverse set of sources, are leading to shifting patterns of both competition and cooperation in developing and deploying the energy technologies of the future.

Our research has shown that six of the large emerging economies (Brazil, Russia, India, China, Mexico, and South Africa, which we refer to as the BRIMCS countries) are now major players in energy technology innovation globally (Atkinson et al., 2009; Kempener, Anadon, & Condor Tarco, 2010; Levi, Economy, O'Neil, & Segal, 2010). These governments and the state enterprises they own invested a minimum of $13.8 billion (purchasing power parity adjusted) in energy research, development, and demonstration (RD&D) in 2008. This is more than the investments of the governments of all of the developed countries represented in the IEA combined, estimated at $12.7 billion for that year. The data on energy innovation investments in the BRIMCS countries are limited and difficult to put in comparable terms, however; there is a clear need to improve data reporting in these countries. China accounts for more than 80% of the BRIMCS total. State-owned enterprises dominate energy innovation in the BRIMCS, amounting to more than 80% of the BRIMCS total (and 90% of China's total). Finally, like the IEA countries, energy RD&D investments by the BRIMCS are concentrated on fossil and nuclear energy. See Table 5.1.

While energy innovation investment in China is surging, the investment in the United States is sinking after the pulse of stimulus spending after the financial crisis. As pointed out by two recent reports (AEIC, 2010; PCAST, 2010), other industrialized countries continue to build on their energy technology expertise, and some of them, such as South Korea and Japan, invest larger fractions of their gross domestic product (GDP) in government energy technology innovation than does the United States

Table 5.1. *Overview of energy RD&D investments in the BRIMCS countries and the United States (Kempener, Anadon, & Condor Tarco, 2010)*

In Million 2008 PPP $Int*	Fossil (incl. CCS)	Nuclear (incl. fusion)	Electricity, transmission, distribution, and storage	Renewable energy sources	Energy efficiency	Energy technologies (not specified)	Total
United States – Gov't	659	770	319	699	525	1,160	4,132
United States – Other	1,162	34	No data	No data	No data	1,350	2,545
Brazil – Gov't	79	8	122	46	46	12	313
Brazil – Other	1,167	No data	No data	No data	No data	184	1,351
Russia – Gov't	20	No data	22	14	25	45	126
Russia – Other	411	No data	No data	No data	No data	508	918
India – Gov't	106	965	35	57	No data	No data	1,163
India – Other	694	No data	No data	No data	No data	No data	694
Mexico – Gov't	140	32	79	No data	No data	No data	252
Mexico –Other	0.1[a]	No data	No data	No data	263[b]	19[c]	282
China – Gov't	6,755	12	No data	No data	136	4,900	11,803
China – Other	289	7	No data	No data	26	985	1,307
South Africa – Gov't	No data	133	No data	No data	No data	9	142
South Africa – Other	164	31[d]	26	7	No data	No data	229
BRIMCS – Gov't	7,100	1,149	≫259	>117	>208	>4,966	>13,799
BRIMCS – Other	2,724	≫38	≫26	≫7	≫289	>1,696	>4,781
BRIMCS – grand total	9,824	>1,187	>285	>124	>497	>6,662	>18,580

* Data from the United States, Brazil, Russia, India, China and South Africa based on 2008, Mexico on 2007.
"Other" includes (whenever available) funding from state and local governments, partially state-owned enterprises, NGOs, and industry.
~ U.S. data on industry expenditure are from 2004 (NSF, 2008).
[a] Based on PEMEX's fund for Scientific and Technological Research on Energy (PEMEX 2008).
[b] Based on 2005 R&D expenditure in car manufacturing industry (CONACYT, 2008).
[c] Based on 2005 R&D expenditure in utilities sector (CONACYT, 2008).
[d] Based on total non-governmental investments into PBMR Ltd. (Department of Public Enterprises, 2010).
> These cumulative values are based on data from only three to four BRIMCS countries, so actual expenditures are likely to be higher.
≫ These cumulative values are based on data from two BRIMCS countries or less, so actual expenditures are expected to be much higher.

(IEA, 2010b). Moreover, the rate of growth of some other countries' investments is far outpacing the U.S. rate of growth. The U.S. government's energy RD&D budget increased 68% between 2000 and 2010 (and has declined somewhat since then), while the comparable budgets of Canada, Finland, Germany, South Korea, and the United Kingdom have expanded 258%, 216%, 104%, 371%, and 669%, respectively (IEA, 2011). Countries such as China and the United Kingdom have also stepped up their efforts to make their investments more effective through the creation of new programs and institutions (Anadon, 2012; Zhi et al., 2013).

5.3. Shifting Patterns of International Energy Innovation Cooperation

These shifting patterns of energy production, use, and innovation investment, combined with a rapidly globalizing world, are provoking changes in the patterns of international cooperation on ERD3. Many of these changes involve tens of thousands of individual researchers and companies working together, often in ways only loosely influenced by any government. There is no requirement for them to report much of this work, and no agency or organization collects comprehensive data. The picture we offer in the text that follows is therefore necessarily a partial glimpse of a set of fast-moving phenomena. Nevertheless, by examining and combining datasets on scientific publishing, technology imports and exports, trade balances, and multilateral and bilateral agreements, a number of patterns can be identified.

5.3.1. Expanding and Shifting Scientific Collaborations

The global science enterprise is growing rapidly, and the share of U.S. authors among all publications in leading scientific journals is slowly decreasing, from 26% in 1999–2003 to 21% between 2004 and 2008 (The Royal Society, 2011). This trend is also evident in scientific publications on energy technology.

To assess these changing patterns, we compiled a dataset of publications in the top peer-reviewed journals covering several areas of energy technology. Figure 5.1 displays an analysis of scientific publications in the top 10 renewable energy journals between 2000 and 2008.[4] The horizontal axis displays the number of publications involving at least one institute from a country, while the vertical axis displays the percentage of publications with international collaborations of a particular country in each year. The chart shows two kinds of data. First, it displays the trajectories the United States, India, South Korea, and China took from 2000 to 2009, with each marker denoting the position of the country in a particular year. Second, the graph shows the position of a selection other developed countries in 2009, including Canada, Finland, Germany, and the United Kingdom, with each country represented by a single

[4] Appendix A5.2.1 in Anadon et al. 2011 describes the methodology behind this analysis.

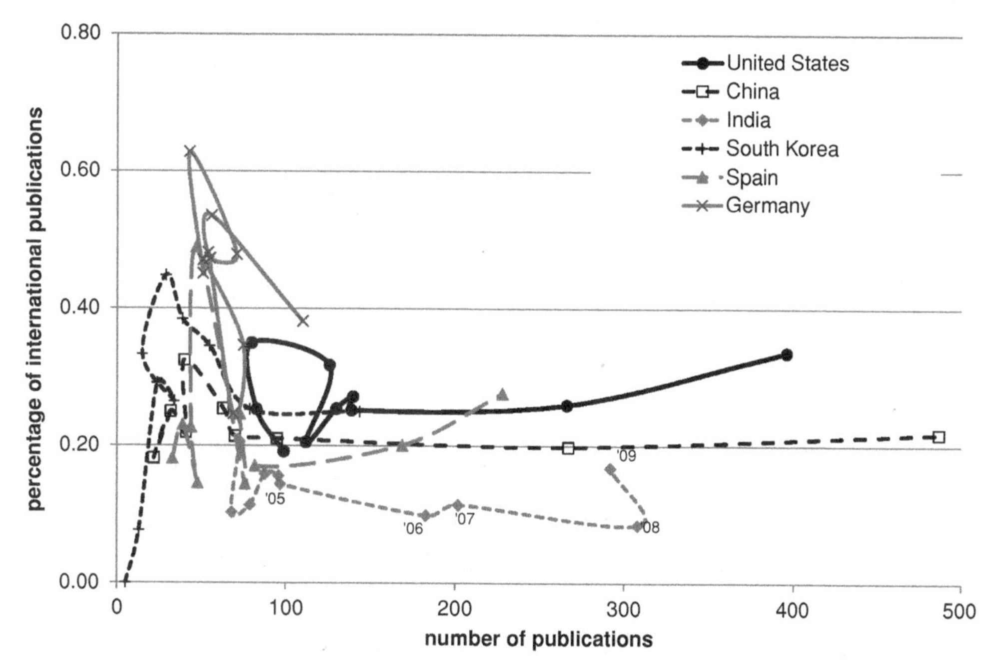

Figure 5.1. International scientific collaborations on renewable energy technologies between 2000 and 2009. The horizontal axis represents the number of publications in the 10 top-ranked renewable energy journals with at least an author from an institution in a particular country; the vertical axis represents the fraction of publications in these journals by institutions from a particular country with at least one author from an institution in a different country.

marker for that year. The United States is included because it is the world's largest producer of scientific literature, including on clean energy technologies; China and India were chosen to represent the growing presence of emerging economies; and South Korea was chosen because of its strategic focus on clean energy technologies. The other countries shown were selected to represent mature developed economies.

As can be seen, all the trajectories are moving strongly to the right – the number of publications is sharply increasing, especially for the United States, China, and India. It also shows that South Korea, starting from almost no collaborations in 2000, has managed to establish itself in the international scientific community, with a number of publications comparable to those of developed countries such as Finland or Germany, and a fraction of international collaborations only modestly lower than those of developed countries. The data suggest that South Korea has established this position through intense collaborations between 2000 and 2005.

These data show that developed countries have a much higher percentage of international collaborations than countries like China and India, although the total output of publications for most modest-sized developed countries is now far behind that of China and India, with their huge populations. Based on these results, and on similar analyses of publications on nuclear and fossil energy technologies, one can conclude that the United States is still a leader in energy technology publications, but that options for international science collaborations are becoming larger and more diverse.

5.3.2. Increasing Import and Export Values for Energy Technologies

Just as patterns of scientific collaboration are changing, the global market for energy technologies is changing as well. In particular, new investments in clean energy technologies increased sixfold between 2004 and 2010, with the highest investment increases over this period in emerging economies such as China, Brazil, and Turkey (PEW Charitable Trusts, 2011).

Furthermore, manufacturing leadership in clean energy technology is shifting from Europe to Asia. In 2012, 86% of the world's solar modules were produced in Asia, with almost two-thirds produced in China alone (REN21, 2013a). Global supply of wind turbines is more broadly dispersed, with four of the top ten producers (who collectively had 77% of the market in 2012) in China, four in Europe, one in the United States, and one in India (REN21, 2013a).

While some countries are insisting on substantial domestic content in their deployments of clean energy technologies, nonetheless it is clear that the value of exports and imports of clean energy technologies around the world is increasing. As just one example, U.S. wind turbine imports soared from $60 million in 2004 to $2.5 billion in 2008 (David, 2009).

More broadly, the monetary value of the import and export of all "power generating machinery and equipment"[5] in both the United States and other countries has been growing steadily since 1990, and the value of the global market for energy technologies continues to increase (UN COMTRADE, 2011). For example, Figure 5.2 shows that from 1989 to 2009 Chinese imports of energy technologies increased by an average of 16% per year, while Chinese exports of these technologies grew 28% per year – though as of 2009 both Chinese exports and Chinese imports in these categories remained far below those of Europe and the United States.

Given the ongoing economic growth in the large emerging economies, the global market for energy technologies is only going to grow, and to continue to shift. Globally, tens of trillions of dollars of energy technology investment are expected over the next two decades, and to the extent that international ERD3 cooperation could result in increased exports of U.S. technologies, the benefits to be gained could be large. For example, a 2008 National Renewable Energy Laboratory (NREL) report estimated

[5] The category "power generating machinery and equipment" (SITC code) includes steam or other generating boilers (SITC 711), steam turbines and other vapor turbines (SITC 712), internal combustion piston engines and parts thereof (SITC 713), nonelectric engines and motors (SITC 714), rotating electric plants and parts thereof (SITC 716), power generating machinery (SITC 718) including nuclear reactors and parts thereof (7187), and wind engines (7189) (UN COMTRADE, 2011).

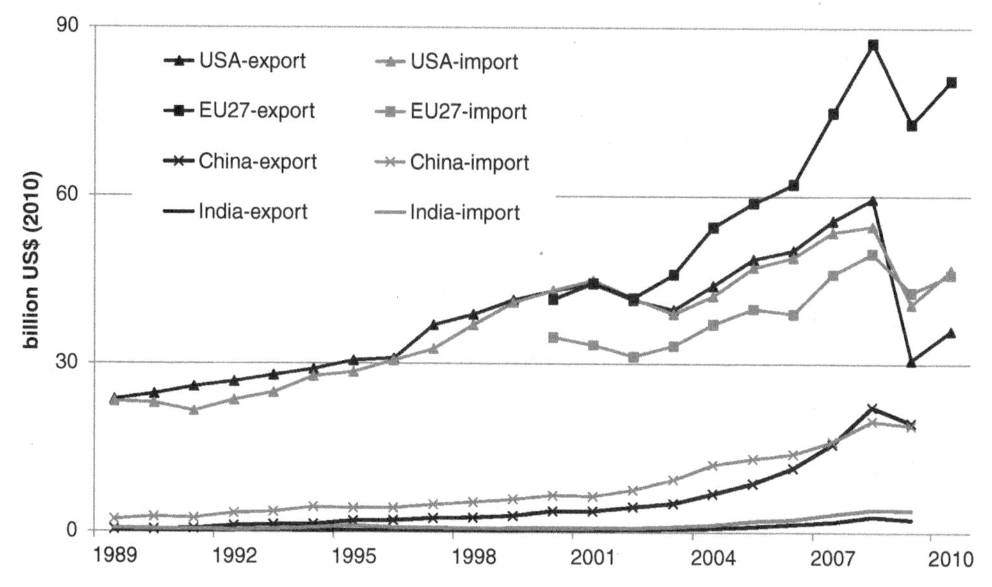

Figure 5.2. Import and export value of "power generating machinery and equipment" (SITC code 71) in the United States, the EU-27, China, and India between 1989 and 2010.

that under certain assumptions, a modest increase and reform of U.S. approaches to international ERD3 cooperation could lead to $40 billion per year in additional exports (NREL, 2008).[6]

5.3.3. Heightened Global Competition in Energy Technology Markets

The United States is facing ever-stiffer competition in global energy markets. China and South Korea, in particular, have clearly made a strategic decision to compete aggressively in the markets for solar, wind, nuclear, and other clean-energy technologies. While the United States must respond to that competitive challenge, the opportunities for cooperation are enormous as well.

Figure 5.2 hints at some of the competitive challenges the United States faces in energy technologies. While the value of China's exports of power generating machinery and equipment increased by 24% per year during 2001–2009, and those of the EU-27 by 10% per year, the U.S. export growth rate in that period was only 1%. These changing global energy technology trade dynamics have direct implications for the U.S. trade balance. Our analysis shows that between 1989 and 1999 the United

[6] These results were derived by estimating the fraction of increased investment in renewable energy and energy efficiency technologies globally ($300 to $600 billion per year by 2020 according to the IEA BLUE Map Scenario and McKinsey [2008]) taking place outside the United States. Subsequently, NREL estimated the fraction of that market that U.S. exporters might be able to gain through increased international cooperation.

States had positive trade balances for all of the main energy technology categories (steam turbines, nuclear reactors, gas turbines, and hydraulic turbines). After 2000, the United States has faced negative trade balances for each of these four energy technology categories (see Anadon et al., 2011, Appendix A5.2.6).

5.3.4. More Bilateral and Multilateral Agreements

Reflecting these trends toward increasing and diversifying global energy technology cooperation, the rate of new bilateral and multilateral agreements on energy technology innovation has substantially increased since 1999. In the decade from 2000–2009, the U.S. government signed 121 new bilateral agreements on ERD3 cooperation (see Anadon et al., 2011, Appendix A5.2). Furthermore, the U.S. government has established two large multilateral frameworks focused in significant part on energy technology innovation (the Asia-Pacific Agreement on Energy and Climate and the Clean Energy Ministerial), and has signed at least 11 other energy technology-specific multilateral agreements.[7] Countries have joined together to establish two new multilateral funds to promote the deployment of clean energy technologies around the world, and since 2007 the United Nations Framework Convention on Climate Change (UNFCCC) has included a separate track on enhanced action on technology development and transfer within the Bali Action Plan (UNFCCC, 2008). As an example of the complexity of these global energy technology agreements, the U.S. government alone is involved in 175 bilateral agreements related to ERD3, and at least 21 international multilateral agreements with an ERD3 component.

5.3.5. Intensified Debate over Intellectual Property Rights and Global Trade Rules

Heightened competition in the global market for clean energy technologies and increased government efforts to support their national energy technology industries have sparked an intensified debate over intellectual property issues (CEPP, 2009; Ockwell et al., 2010) and global trade rules (Huang, 2011). On the one hand, policymakers in Western countries are concerned about the lack of enforcement of intellectual property rights in emerging economies despite international trade rules protecting such rights (Reuters, 2008). Some policymakers in developing countries, however, have pushed for financial assistance or other help in gaining the capability and the permission to produce new energy technologies, arguing that the high costs charged by developed countries for new energy technologies are a key barrier to technology transfer (UNFCCC, 2006).

[7] Much of the technological cooperation work of the Clean Energy Ministerial is proceeding under the 21st Century Power Partnership (http://www.21stcenturypower.org/).

In contrast to the political discourse around intellectual property rights (IPR) issues, recent empirical work has found that IPR controls have not posed a major barrier to the spread of clean energy technologies, and that concerns over theft of intellectual property in clean energy are largely overblown (Sagar & Anadon, 2009; Gallagher, 2014). The international business literature does not see IPR issues as a barrier to international cooperation, but rather as an important administrative feature of inter-firm relations that needs to be resolved at the start of any joint venture, subsidiary arrangement, or foreign direct investment. The business literature emphasizes unfamiliarity with foreign legislation, cultural differences, lack of market knowledge, communication problems, and lack of trust as the important barriers to international innovation activities (Beamish & Killing, 1996; von Zedtwitz & Gassmann, 2002). Similarly, in the scientific community the lack of trust in foreign scientific capabilities is more difficult to resolve than intellectual property issues, for which there have been established several good practices (Kasprzycki, Ozegalska-Trybalska, & Mayr, 2008).

Regardless of IPR issues, the increasingly competitive environment for energy technologies is creating mistrust at a government level, with disputes over what efforts constitute reasonable support for national ERD3 activities and what constitute unfair subsidies. In 2007, Brazil and Canada challenged the U.S. subsidies for corn and other agricultural projects in the context of biofuels, and Brazil challenged the U.S. subsidies for domestic ethanol use under the World Trade Organization (WTO) trade rules. In addition, the United States long had a tariff on Brazilian ethanol. In 2009, the European Union (EU) complained about Canada's requirement for domestic content for its renewables feed-in tariff, and in 2011 the United States challenged China's export subsidies for its domestic wind turbine companies (Huang, 2011). More recently, both the United States and the EU have accused China of dumping solar panels on the world market for less than their full cost. At the same time, charges that Chinese hackers are stealing U.S. firms' commercial secrets through cyber intrusions – including the secrets of clean energy firms – have become a hot political issue in the United States.

5.4. International Energy Innovation Cooperation: What the U.S. Government Does So Far

5.4.1. Current U.S. Government ERD3 Cooperation Initiatives

U.S. government initiatives on international ERD3 cooperation efforts have a long history, from nuclear energy cooperation in the 1950s to the first comprehensive multilateral frameworks developed after the 1973 oil shock to today's Clean Energy Ministerial (CEM). Many of the initiatives by the Obama administration are a continuation of the bilateral and multilateral initiatives established by the Bush administration, although they have intensified in two ways: an increased focus on implementation and intensified interagency coordination.

A primary example of the intensification of international ERD3 cooperation is the Memorandum of Understanding (MOU) between the United States and China signed in July 2009, which elaborates on earlier agreements signed by President Bush.[8] In November 2009, President Barack Obama and President Hu Jintao announced seven U.S.–China clean energy initiatives that the United States and China would carry out under the terms of the MOU.[9] Each of these initiatives has been translated into separate programs and projects. One of these initiatives resulted in the creation of a U.S.–China Clean Energy Research Center, to which both countries are contributing $150 million over 5 years from public and private sources. The Center includes three consortia on "Building Energy Efficiency," "Clean Vehicles," and "Advanced Coal Technology," each consortium involving a mixture of U.S. and Chinese universities, national laboratories, departments, and companies (see CERC, 2013 for a discussion of the progress made so far in each of these areas). Besides this Center, the agreement has resulted in activities ranging from renewable energy workshops and studies to shale gas assessments to collaborative experiments on coal gasification (DOE, 2011). A range of commercial energy technology deals were established during the Chinese state visit to the United States in January 2011 involving a large number of U.S. companies organized under the U.S.–China Energy Cooperation Program.[10] For example: Duke Energy and the ENN Group signed an MOU to cooperate in a demonstration project named "Eco-City" to test clean coal, electric vehicles, and energy efficient buildings in Langfang (China); American Electric Power (AEP) signed cooperation agreements with three companies – China Huaneng, State Grid Corporation of China, and China National Offshore Oil Corporation (CNOOC). A similar intensification process has taken place in the United States–India bilateral agreement (described in more detail in Anadon et al., Appendix A5.3).

Interagency coordination has also become more important in the Obama administration.[11] The Director of the Office of Science and Technology Policy (OSTP) filled the position of associate director for National Security and International Affairs, responsible for overseeing U.S. science and technology cooperation with other countries, which was vacant during the Bush administration (Stine, 2009). Specific to

[8] In June 2008, the U.S.–China Ten Year Framework for Cooperation on Energy and the Environment (TYF) was established with task forces for (1) clean, efficient, and secure electricity production and transmission; (2) clean water; (3) clean air; (4) clean and efficient transportation; and (5) conservation of forest and wetland ecosystems (Office of Public Affairs, 2008).

[9] The seven initiatives included the initial five task forces established under the Ten Year Framework and two additional task forces on "energy efficiency" and "protected areas/nature reserves."

[10] www.uschinaecp.org/.

[11] The National Science Board highlighted the lack of coordination of international science and technology partnerships in a report in 2001 (NSB, 2001). This report was followed up by a report in 2008 that again called for a "coherent and integrated U.S. international science and engineering strategy" due to "shifts in the international landscape, along with the unfulfilled recommendations of its 2001 report" (NSB, 2008).

energy, the administration has set up two interagency working groups, one on biofuels and one on carbon capture and storage, and both groups have stressed the need to coordinate efforts to ensure maximum effectiveness of international cooperation activities (Interagency Task Force on Carbon Capture and Storage, 2010; IWG, 2010).[12]

5.4.2. Cataloging and Categorizing the Elements of International ERD3 Cooperation

The new international ERD3 cooperation initiatives introduced by the Obama administration are a positive development, but also highlight the complexity of international ERD3 cooperation. The complexity in which decision making on international cooperation takes place includes:

- Multiple *platforms* for international cooperation that are operating simultaneously
- The involvement of a large number of different *stakeholders*, ranging from the White House to departments to regulatory agencies to national laboratories to industry
- Different *channels* through which cooperation takes place (i.e., government-to-government; university-to-university; industry-to-industry; or mixed), often involving different sets of actors
- Multiple and overlapping *activities* within each platform or channel, which are subsequently divided into smaller projects.

Most international cooperation includes combinations of different platforms, channels, stakeholders, and activities. To understand and to analyze international cooperation projects, the connections between these different elements need to be considered. Take the case of the long-standing international cooperation on fusion energy technologies, the International Thermonuclear Experimental Reactor (ITER). The proposal for ITER was made during the end of the Cold War, and had clear geopolitical significance for the signatories at the time. Nowadays, ITER benefits from established national agencies that are responsible for the national coordination of activities and/or tenders for industry. Program officers involved in these activities often have different concerns than those that operate at a geopolitical level, such as how to maintain funding for the ITER project while its project costs are rising. Finally, there are the scientists concerned with the practicalities of solving the workings of nuclear fusion. At this level, concerns might arise around the number of training places, the scientific contributions of the different teams, and other risks associated with international scientific projects.

Elsewhere, we have catalogued and categorized the very large number of agreements and organizations focused on international ERD3 cooperation in which the

[12] A similar interagency task force for Hydrogen and Fuel Cells was established by the Bush administration after EPAct 2005.

United States participates (Anadon et al., 2011, Chapter 5 and Appendix A5.1.3). That examination makes clear that U.S. cooperation with many countries proceeds under the aegis of many different agreements. Even a single technological area, such as solar technology or nuclear fission, may be included in multiple agreements in which a particular foreign country participates.

Our examination of the existing U.S. international ERD3 agreements and initiatives also makes clear that most international ERD3 cooperation projects supported by the U.S. government are deployment activities. Rather than developing new technologies, they focus on the development of databases, handbooks, best practice examples, state-of-the-art literature reviews, standards and codes, computer models, regulatory issues, or policy reform advice. This is to be expected, because generally these activities are cheaper and easier to arrange than projects in which equipment is being purchased and new processes are being developed. There are several international cooperation projects that support the exchange of experimental data, collaborative testing of materials, or the coordination of experimental testing in facilities. Only a small number of international cooperation projects actually work on the development of new energy technologies, and these projects mainly concern fusion and fission energy technologies.

5.4.3. U.S. Agencies Involved in International ERD3 Cooperation

Agreements are often led by a single U.S. agency, but there is often more than one agency involved in implementation. For example, the Asia-Pacific Partnership on Clean Development and Climate (APP) was led on the U.S. side by the State Department but included six task forces (each of which can be considered a separate program), each coordinated by one or more government agencies. Similarly, the CEM is led by the Department of Energy (DOE) but has 11 initiatives (programs), and each initiative involves different U.S. agencies. Overall, there are at least nine U.S. government departments and ten U.S. agencies involved in different international ERD3 cooperation programs.

Tables 5.2 and 5.3 provide an overview of the main departments and sub-cabinet agencies involved in international cooperation on energy technology innovation and their roles. Several departments and agencies have programs that affect the same innovation stages (R&D, demonstration, market formation, and deployment), and the same energy technologies (in particular, renewable energy and energy efficiency technologies). In addition, in many cases, agencies and departments also have common purposes for engaging in international cooperation (economic, environmental, developmental, and political). However, each department and agency has different organizational and administrative structures, different finance mechanisms at their disposal, different channels of involvement with industry and scientists, and different levels of engagement with Congress and within the White House.

Table 5.2. *Overview of U.S. departments and their role in international ERD3 cooperation*

U.S. Departments	Role in international ERD3 cooperation
Department of the Treasury (Treasury)	Treasury administers funds for the Clean Technology Fund (approved by the World Bank and funded by multilateral development banks), the Strategic Climate Fund (including the program for Scaling Up Renewable Energy in Low-Income Countries, known as SREP), and the Global Environment Facility (a joint project of the World Bank and UNEP).
Department of State (DOS)	DOS's Bureau of Oceans and International Environmental and Scientific Affairs (OES) is responsible for reporting its Climate Action Plans to the UNFCCC and for managing several international agreements. Within OES, the Office of Science and Technology Cooperation (OES/STC) is responsible for all legally binding international agreements on science and technology. The Office of International Energy and Commodity Policy (EC) is responsible for U.S. international energy policy. Finally, State manages the budget for the United States Agency for International Development (USAID).
Department of Energy (DOE)	DOE has the largest direct impact on international energy technology cooperation with activities taking place within its technology programs, program offices, and the national laboratories. DOE has less funding for international activities than State or Treasury.
Department of Commerce (DOC)	DOC's International Trade Administration (OTA) works to promote U.S. exports of energy and environmental technologies; provides information on and organizes events about fast-growing energy markets around the world; the National Institute of Standards (NIST) is involved in developing international energy technology standards such as internationally compatible biofuels standards (Chum, 2009); and the National Oceanic and Atmospheric Administration (NOAA) indirectly affects energy technology deployment through its international science collaborations collecting and analyzing international air and water quality data (DOC, 2010a). DOC has also set up advisory committees to help them craft policies promoting the export of renewable energy and energy efficiency technologies (DOC, 2010b).
Department of Transportation (DOT)	DOT is involved in bilateral and multilateral agreements on vehicle technologies and climate change issues; international monitoring of environmental impacts of energy technologies; international efforts to curtail environmental impacts of transportation; and policy support for foreign governments.
Department of Defense (DOD)	DOD conducts international collaborative R&D projects and is responsible for energy infrastructure development in countries such as Iraq and Afghanistan.

(continued)

Table 5.2 *(continued)*

U.S. Departments	Role in international ERD3 cooperation
Department of Agriculture (USDA)	USDA is involved in setting international standards for biofuels; science and technology collaborations on sustainable and renewable bioenergy technologies and use; and partnerships with the interagency organizations for climate change technology policy (CCTP) and climate change research (USGCRP).
Department of Interior (DOI)	DOI coordinates projects in the APP; is a partner of CCTP, USGCRP, and USAID; and administers the International Technical Assistance Program, which supports projects preventing the decline of environmental quality through science and technology cooperation and aid.
Department of Housing and Urban Development (HUD)	HUD designs and manages bilateral and multilateral information exchange programs on housing and urban policies; and partners with USAID, State, American Embassies, CCTP, and USGCRP.

As we discuss in the text that follows, with so many countries, technologies, stakeholders, and initiatives involved, it is effectively impossible to coordinate and prioritize everything at once. It is also very difficult to collect and analyze the data to make it possible to learn by doing about what works and what does not, under what circumstances. It is important to recognize, however, that overlap in activities is unavoidable and cannot (and should not) be resolved through centralized management. Because of the uncertain nature of innovation, and the fact that good ideas arise in unpredictable ways, letting many flowers bloom – even if some ultimately bear little fruit – is the best way to proceed.

5.4.4. Activities in Government International ERD3 Cooperation

Three different kinds of international cooperation activities can be distinguished. The project level is where real progress in developing new energy technologies takes place – but it is also the level where there is the least publicly available aggregate information about who is involved, what activities are going on, and what financial and technical resources the projects are using. Projects are often coordinated and managed by sub-organizations within U.S. departments or agencies. Within DOE, the 21 national laboratories, containing more than 30,000 scientists and engineers, play a very important role in coordinating and carrying out projects on international ERD3 cooperation. For example, the China Energy Group and the International Energy Studies Group at Lawrence Berkeley National Laboratory, the U.S.–Brazil agreements coordinated at NREL, the International Group at the National Energy Technology

Table 5.3. *Overview of government agencies directly involved in international ERD3 cooperation (continued)*

U.S. Agency	Role in international ERD3 cooperation
United States Agency for International Development (USAID)	Bilateral agreements to support foreign countries in power sector reforms, developing energy access, and deployment of clean energy projects reducing greenhouse gas emissions.
United States Trade and Development Agency (USTDA)	Bilateral agreements, trade missions, and financial support encouraging export of U.S. energy technologies. Provides funding for energy-related projects in foreign countries.
United States Export-Import Bank (Ex-Im)	Trade missions and financial support (insurance, working capital, loan guarantee programs) encouraging export of U.S. energy technologies. Environmental Exports Program (EEP) provides enhanced financial support for renewable energy, clean coal, energy efficiency, and fuel cells (State, 2010).
Overseas Private Investment Corporation (OPIC)	Provides project financing and political risk insurance for U.S. companies, including independent power projects, and supports private investment funds in emerging economies.
Environmental Protection Agency (EPA)	Involvement in bilateral and multilateral agreements on climate change issues, international monitoring of environmental impacts of energy technologies, international efforts to curtail environmental impacts, and policy support for foreign governments and regulators.
Office of Science and Technology Policy (OSTP)	OSTP coordinates U.S. international technology cooperation; engages directly with technical experts and leadership teams in different federal agencies; and discusses international cooperation opportunities with its counterparts in other countries.
Council on Environmental Quality (CEQ)	CEQ has a domestic role coordinating federal environmental efforts. Through its initiatives on the Interagency Carbon Capture and Storage Task Force and the Interagency Climate Change Adaptation Task Force, CEQ has some involvement in international cooperation.
National Aeronautics and Space Administration (NASA)	International science collaborations on energy technologies, such as space solar energy, and support for deployment of energy technology through its Earth Observation Data.

(continued)

Table 5.3 *(continued)*

U.S. Agency	Role in international ERD3 cooperation
National Science Foundation (NSF)	Grants for international science collaborations on energy technologies. In the 1970s NSF had. specific grants for "international energy technology cooperation," but current grants are either energy-specific or country-specific.
U.S. Embassies	Science and technology staff at U.S. embassies provide information and help facilitate international cooperation activities. However, the State Department has shifted from science and technology counselors at major embassies to more junior staff. DOE now has offices in embassies in 13 countries, including Russia, Japan, France, China, and India: these DOE offices play an important role in collecting information on energy developments in these countries and facilitating cooperation.

Laboratory, and the U.S. ITER team at Oak Ridge National Laboratory all play a central role in managing, executing, and coordinating international cooperation activities.

Many individuals and groups within these national laboratories also run and manage many of the bilateral and multilateral agreements signed by the U.S. government. National laboratories not only are involved in managing projects derived from programs within these agreements, but they are also a major source for international cooperation activities themselves. For example, our analysis of scientific collaborations has shown that NREL has about 30% of its publications in collaboration with research institutions outside of the United States.[13] With increasing international ERD3 cooperation activities, the number of non-government entities involved in U.S. government-supported international ERD3 activities increases as well. The U.S. government engages the private sector, state and local governments, and nonprofit organizations in both the management of and as participants in international ERD3 cooperation, while at the same time these actors are increasingly engaging in international ERD3 activities themselves.[14] Traditionally, there has been only modest involvement of private companies in such activities, but that involvement is growing; the consortia

[13] From 1995 to 2009, between 27% and 31% of NREL's publications were in collaboration with foreign institutions. Our analysis of renewable energy technology papers reveals that U.S. research institutions publish on average 15% of their publications with foreign institutions in 1995, 19% in 2000, 25% in 2005, and 34% in 2009.

[14] The R20, an international organization connecting states, local governments, and cities globally, is one of those examples whereby non-federal government actors initiate international ERD3 activities.

focused on particular areas of energy technology in the U.S.–India and U.S.–China energy centers described previously, for example, include private firms along with public entities.

5.4.5. Coordinating ERD3 Cooperation within the U.S. Government

There is no single process or structure for coordinating international ERD3 cooperation within the U.S. government, and no overall strategic plan for such efforts. Instead, there are interagency groups that coordinate portions of these efforts, and a sophisticated network of personal contacts, both within the different U.S. agencies and across agencies, through which coordination of international cooperation activities takes place.

For example, DOE has established a series of briefing discussions where it informs policymakers from different U.S. agencies of progress within its programs. However, this network depends on a small number of highly connected individuals. Furthermore, this system of coordinating international cooperation activities is not replicable (every time there are new opportunities for international cooperation, policymakers will have to go through the effort of collecting information through their personal networks); its knowledge is tacit; and its information is not easily accessible to private firms, university or laboratory researchers, or NGOs.

The picture of coordination on these topics within the federal government includes both progress and challenges.

- At the federal level, the "Green Cabinet," including the Secretaries of Energy, Interior, Agriculture, Transportation; Housing and Urban Development; and Labor; the Administrator of the Environmental Protection Agency; the Administrator of the Small Business Administration; the head of the Council on Environmental Quality; the director of the Office of Science and Technology Policy; and others, offers an opportunity for high-level coordination of energy and environment initiatives, including international ERD3 cooperation.[15] At this high level, however, only the most important and urgent issues are likely to be addressed.
- As noted earlier, the Obama administration has expanded the emphasis on interagency coordination of international ERD3 cooperation at the working level, creating new interagency groups for that purpose.
- DOE's Office of International Affairs (IA) coordinates many of DOE's international ERD3 activities, working with officials from the R&D offices themselves. That office also coordinates with other agencies. It has no mandate or authority to control the ERD3 activities of other agencies, however, some of which are important to international ERD3 progress. The IA also collects and disseminates information about DOE's international commitments.

[15] Originally, Carol Browner, director of the White House Office of Environment and Climate Change, chaired the Green Cabinet. That office was phased out in 2011, with its duties folded into the broader Domestic Policy Council.

- The Renewable Energy and Energy Efficiency Export Initiative established in 2010 spans eight federal agencies and involves a coordinated action plan, which consists of developing new financing products and investment vehicles; addressing trade barriers; creating new export opportunities; ranking new markets; increasing the exposure of foreign buyers to U.S. renewable energy and energy efficiency technologies; and providing U.S. companies with better information on U.S. government export programs (TPCC, 2010). The Obama administration has established similar support for nuclear reactor exports, and other initiatives help support exports related to fossil technologies.
- The Council on Environmental Quality (CEQ) and the Office of Science and Technology Policy (OSTP) can establish formal guidelines and priorities for international cooperation. But they do not control the budgets of any of the agencies that would be implementing ERD3 cooperation activities, and they do not have the necessary staff or the mandate to oversee the actual implementation of the myriad international ERD3 activities underway. (CEQ and OSTP do, however, influence budget decisions made by the Office of Management and Budget.)
- The Department of State provides support for international cooperation (both through its headquarters staff and through embassies around the world). But it is neither equipped for nor responsible for taking the lead in advancing energy technology innovation goals.
- The national laboratories, such as NREL and Lawrence Berkeley Laboratory, coordinate many DOE-led international cooperation projects. For example, in 2004, NREL coordinated projects involving 5 different political objectives, 14 different activities, and 5 different groups of stakeholders (Taylor, 2004). But they do not have the mandate to coordinate international cooperation activities across federal agencies.
- The Climate Change Technology Program (CCTP) and the United States Global Change Research Program (USGCRP)[16] attempt to coordinate a subset of energy technology-related activities across the government. The CCTP supports other departments in their initiatives and reports on their international cooperation activities, but traditionally has not had the means to coordinate international cooperation activities across agencies (CCTP, 2006, 2009). USGCRP has an Interagency Working Group on International Research and Cooperation[17] that coordinates the inputs into major scientific international efforts, such as the Intergovernmental Panel on Climate Change (IPCC) or Inter-American Institute for Global Change Research (IAI) (USGCRP, 2009). However, the main focus is on climate change research and the measurement of global environment change rather than on energy technology RD&D.
- In terms of information the public could use to understand the effort, the Open Government Directive, which requires executive departments and agencies to publish government information online, resulted in the creation of OpenEI,[18] an online platform that publishes information about U.S. government activities on energy. OpenEI also publishes information concerning existing and new international cooperation activities, but the data do not allow for systematic comparison of activities.

[16] Formally known as the United States Climate Change Science Program (USCCSP).

[17] The Working Group includes representatives from USDA, DOC, DOD, DOE, HHS, DOI, State, DOT, USAID, EPA, NASA, NSF, and the Smithsonian Institute.

[18] http://en.openei.org.

5.4.6. Summary of Current State of Government-Supported International ERD3 Cooperation

The U.S. government is supporting a broad range of international ERD3 cooperation, involving many technologies, many partners, multiple stages of technology development, and many U.S. government and non-government participants. In many cases, this cooperation has led to significant progress that has served U.S. interests well.

But as with other areas of energy technology innovation discussed in this volume, the pace and scale of these efforts do not match either what is likely to be required to meet this century's energy challenges or the opportunities available. More needs to be done – and done with an increased focus on performance in each project.

The increase in the number of platforms, agencies, and activities has created more options and opportunities for international ERD3 cooperation, but it has also created new challenges for coordination and prioritization. The U.S. political environment, discussed in Chapter 1, has also contributed to uncertain funding and shifting priorities, making it nearly impossible to set consistent strategic directions for these efforts. Combined, these conditions have led to:

- Little continuation and follow-up between activities
- Duplication and repetition of activities supported across the government
- Restructuring of existing projects with little evaluation
- A large number of inactive agreements
- Limited efforts to extract lessons from international cooperation activities and to provide them to the wider community

5.5. International Energy Innovation Cooperation: The Challenge of Choice

The complexities and challenges of coordinating international ERD3 activities are clear. The U.S. government faces several fundamental conundrums as it seeks to coordinate and prioritize its efforts in international ERD3 cooperation:

1. The U.S. government has many objectives it is pursuing through international ERD3 cooperation, which complement each other in some cases and contradict each other in others. These include, among others, mitigating climate change, promoting U.S. exports, reducing U.S. and world dependence on oil and gas from volatile regions, strengthening political relations with key countries, and alleviating energy poverty.
2. A huge number of activities are underway, involving many countries, many technologies, many different types of activities, and many different stakeholders.
3. Many different parts of the U.S. government are pursuing this work, with sometimes complementary and sometimes conflicting interests and priorities.
4. The U.S. government has little control over (and incomplete information about) many of the most important cooperative activities taking place, driven by private firms, universities, individuals, and teams.

5. Major progress on a large number of technological fronts is likely to be needed to meet the energy challenges the United States faces.
6. No comprehensive, centralized source of information is available that includes all of these activities, and no one is collecting the kinds of data that would allow systematic learning by experience in how best to implement these programs.
7. The results of R&D are inherently uncertain. No one knows which projects will lead to breakthroughs and which will lead to little or no progress. Even after the fact, measuring the impacts of international ERD3 activities is very difficult. Unintended consequences, uncertainties, and the long-term nature of the innovation process are some of the factors inhibiting simple comparison or prioritization between activities.
8. Balancing competition and cooperation in this complex, multifaceted, and fast-changing environment is inherently difficult.
9. Cooperation activities take (at least) two parties. The U.S. government will need to consider the interests and values of partnering governments.
10. International cooperation activities do not start from a clean slate. Historical relations, or the conditions under which new activities are initiated, often constrain the range of choices or the implementation of activities.

In trying to resolve these challenges, the U.S. government can draw from the experience of private firms, which have long been seeking to find the best approaches to international cooperation and competition to maximize their risk-adjusted profits (Ronstadt, 1978; Patel & Vega, 1999; Belderbos, Kyoki, & Iwasa, 2009). However, even within the single overriding metric of maximizing profit, complexities, trade-offs, and preferences arise: is it better to pursue projects that pose little risk but offer only moderate payoffs, or is it better to pursue high-risk efforts with the potential for large returns? How does a firm identify cooperative efforts that offer substantial net benefits both to the firm and to its potential partners – without which the projects are unlikely to move forward? For the U.S. government, the complexities of making the judgment are far greater, because the U.S. government has many different goals that might be served by cooperative ERD3 activities, many different stakeholders pursuing these activities, and pursues these activities through many different platforms and pathways.

Similarly, there is a considerable amount of literature on how "national systems of innovation" are globally connected (Edquist, 1997; Archibugi & Iammarino, 1999; Archibugi & Iammarino, 2002). However, strategies for international cooperation go beyond the establishment of policies that support national companies operating in global sectors. Furthermore, international cooperation often has political implications that go beyond the value that such activities might have for national innovation activities (Haas, 1975; Putnam, 1988; Young, 1989; Iida, 1993). Given this complexity, how can the U.S. government improve its efforts to choose, plan, coordinate, and learn from cooperative international ERD3 activities? Here, we focus first on choosing which efforts to pursue, starting with existing studies that have proposed criteria for

Table 5.4. *Strategic criteria for areas (technology and country choices) for international ERD3 cooperation projects*

Interagency Task Force priority ranking (1973)	PCAST strategic criteria (1999)	NREL criteria for country and technology choices (2008)
1. Existence of unexploited opportunities 2. Existence of useful technology abroad 3. Potential for reduction of energy deficit 4. Transfer time to commercial use 5. Lack of legal or proprietary barriers	1. Reducing consumer costs and promoting economic benefit 2. Increasing access to foreign markets 3. Reducing local and regional air pollution 4. Reducing of global greenhouse gas emissions 5. Dampening growth of world oil demand 6. Improving proliferation resistance of civilian nuclear energy system	1. Large clean energy markets 2. Commitment to growth in dean energy markets 3. R&D centers with significant budgets 4. Effective IP protection 5. Strong relationship with the United States 6. Global or regional leader

Based on Polack & Congdon (1974); PCAST (1999); and NREL (2008).

selecting priority areas for international ERD3 cooperation, principles for designing specific projects, and a range of metrics for evaluating the impact of projects (Pollack & Congdon, 1974; PCAST, 1999; IEA, 2005; NREL, 2008; Asia Society, 2009; Bazilian et al., 2009). We then compare these proposals with the approaches the U.S. government is now using to manage international ERD3 cooperation and offer suggestions for improvement.

5.5.1. Decision Criteria for International ERD3 Cooperation

Over the years, several groups have proposed criteria for prioritizing which countries and technologies should be the focus of ERD3 cooperation efforts. Table 5.4 provides an overview of three sets of criteria that have been proposed. The criteria put forward by the Interagency Task Force on "International Cooperation in Energy Research and Development" were specifically created to support priority ranking (Pollack & Congdon, 1974), while the NREL criteria were proposed for selecting potential countries and technology partnerships (NREL, 2008). In contrast, the PCAST strategic criteria were developed to describe the space or "overall portfolio" of international cooperation activities on energy technology innovation with separate "project criteria" and "public–private interface criteria" to choose among projects within this space (PCAST, 1999). Although there is some overlap between the different proposed criteria, there are also major differences, reflecting the diversity of policy drivers for international ERD3 cooperation, and how these drivers have shifted and expanded over time.[19]

The reality is that no single set of criteria can – or should – drive U.S. decisions on all of its international ERD3 cooperation activities. These activities are pursuing objectives ranging from alleviating energy poverty to reducing greenhouse gas emissions to reducing nuclear proliferation risks, and each of these are legitimate U.S. interests. It is legitimate and appropriate that different international ERD3 efforts should be guided by different criteria and measures of success.

Nevertheless, it is clear that a few of the U.S. objectives are high priorities of the U.S. government. Opportunities to cooperate with foreign countries should be pursued if analysis suggests that they offer strong potential for:

- Contributing to technologies that have the potential to play a substantial role in mitigating climate change
- Expanding the likely future share of jobs in energy technology markets that are in the United States
- Reducing the security and economic costs of undue reliance on volatile oil and gas markets
- Reducing energy costs to U.S. consumers
- Reducing the risk of nuclear proliferation

The State Department's efforts provide an example of how goals beyond energy technology innovation shape international cooperation. The State Department structures cooperation agreements around the potential "ends" that they provide, such as poverty alleviation, rural development, or climate change, and sees technology choices only as a "means" to achieve these ends. In contrast, agencies such as the Trade and Development Agency (TDA) or the Commerce Department make cooperation choices to support the transfer of specific technologies, and are less concerned about the potential wider implications of these technologies within different countries. Even within one government agency, several offices might not have the same agenda. For example, the energy and climate change bureaus at the State Department both work on energy issues within an international context, but they have different policy objectives.

In short, because the United States has many objectives, it cannot use one set of criteria to drive all of its ERD3 cooperation; but on the other hand, there is a strong need to set some form of priorities. In response to this situation, we recommend

[19] Our historic analysis of policy documents on international ERD3 cooperation shows at least eight different political drivers that, at different points throughout history, have shaped the policy debate. These drivers are: (1) accessing knowledge on high-tech energy technologies; (2) retaining or gaining technological competitiveness; (3) increasing energy security; (4) accelerating the commercialization of new energy technologies; (5) reducing the negative environmental impacts of the production and consumption of energy; (6) promoting reforms in the energy sector; (7) providing energy access to the poor; and (8) increasing U.S. access to emerging markets. None of these drivers has become less important. Instead, this amalgamation of different historical and contemporary drivers for international ERD3 cooperation has led to a more complex situation for U.S. decision makers in their process of selecting *areas* of international ERD3 cooperation.

the use of heuristics to aid decision making and the implementation of a two-track approach to international ERD3 cooperation, encouraging both bottom-up initiatives and top-down direction of a few key strategic priorities (see Section 5.6).

5.5.2. Designing International ERD3 Cooperation Activities

In addition to identifying the priority countries and technology areas for collaboration that are of strategic interest to the United States, policymakers in the U.S. government must make choices about the kinds of activities in which they should engage; namely, what the portfolio of activities should look like, what financial or human resources should be invested, which partners should be involved, and what the timescales and objectives of the projects should be.

The PCAST study of 1999 recommended a portfolio approach toward international ERD3 activities, with programs and projects spanning across:

- The timescales of the activities: near-term (<5 years), medium-term (5–20 years), and long-term (>20 years)
- The different technological pathways that government activities support[20]
- The stage of innovation: R&D, demonstration, market formation, and deployment
- The participants involved: local and national governments, the private sector, local, national, and international NGOs, multilateral development banks, and other international institutions

Overall, it appears that the U.S. government has adopted such a portfolio approach across multiple dimensions – whether in response to the PCAST recommendations or not. The U.S. government supports projects covering a diverse range of time frames, technological pathways, and innovation stages (though its support for demonstration projects is somewhat limited) – and many participants are involved. Thus the U.S. government's ERD3 cooperation activities do reasonably well on the first of the CASCADES criteria for an effective energy innovation policy outlined in Chapter 1– taking a comprehensive approach. Nevertheless, it is impossible to do everything; there is a need for clear criteria for choice.

5.5.3. Structural versus Procedural Criteria for Designing Cooperation Projects

Table 5.5 provides two sets of criteria developed by PCAST (1999) and the IEA that these groups offered to policymakers for designing specific ERD3 cooperation

[20] The term "technological pathway" instead of "technology area" was defined within the PCAST (1997) report on federal energy research and development to reflect that the notion that multiple pathways can occur within a particular technology area.

Table 5.5. *Criteria to consider when designing specific international cooperation projects*

PCAST project criteria (1999)	IEA Implementing Agreement criteria (2010)
• Mutual interests, evident to all participants • Mutual investments demonstrating commitment and common standing to participants	• Determine shared objectives and mutual benefits. • Demonstrate mutual commitment of partners. • Develop equitable funding, sharing, or subsidy arrangements.
• Agreements on goals and milestones, timetables, responsibilities, sharing of intellectual property rights	• Set implicit and explicit goals. • Identify "champions" if required. • Set parameters of rights and responsibilities. • Develop procedures for bringing in new partners or opting out of existing partners. • Establish intellectual property and patent rights.
• Streamlined management, including clear lines of communication	• Develop common vision and strategic intent. • Delineate tasks with skills and resources required. • Develop internal and external communication.
• Continuity of support matched to timeline and milestones • Periodic independent oversight	• Set initial term and procedures for extension and termination. • Set milestones and metrics for effectiveness. • Develop tracking and evaluation procedures.
• Structural readiness	• N/A

projects. Both sets of criteria describe the structural features of an international cooperation project, but not the procedural steps that create such features.

An analysis of the IEA Implementing Agreements – its primary network for international ERD3 cooperation among its member countries – suggests that the structural features of international ERD3 projects generally adhere to these design criteria (see Anadon et al., 2001, Appendix A5.2.3).

Most annexes start with a description of why the project is of interest to the participants, define their "purpose and objective," identify an "operating agent" that administers the project, define subtasks and assign responsibilities, and provide a timeline for progress and milestones. For example, the IEA Implementing Agreement on Hydrogen provides such descriptions for 9 out of its 11 currently operating projects. Similarly, State's handbook on drafting international science and technology agreements specifies that each agreement needs to include a description of the scope; funding; joint committees and councils; clauses about intellectual property allocation; and a description of the duration, amendment, and termination of agreements (State, 2001).

Academic literature on interorganizational relationships, however, stresses that the ultimate form of the project, as defined by its objectives, goals, and milestones, is often less important than the personal relationships and tacit understandings that are created in the time before any project comes to paper. These initial relationships are often governed by a range of short-term, ad hoc efforts (Van de Ven & Gordon, 1984), and it is important to provide sufficient support to allow these personal relationships to grow. Especially in international cooperation activities, these so-called procedural aspects are often more important than the specific text on which the parties ultimately agree, because personal relationships and tacit understandings are more effective in mitigating any risks and uncertainty associated with interorganizational collaborations (Parkhe, 1993; Das & Teng, 1998). This is particularly relevant in the context of international ERD3 cooperation, where lack of information, cultural differences, communication problems, and the unpredictable nature of innovation itself create high levels of uncertainties.

Our interview results confirm the importance of the procedural aspects in creating effective international ERD3 cooperation projects. Both scholars involved in international cooperation projects and project managers at DOE mentioned that the key to successful projects is the organization of effective meetings in which people (scientists, technologists, companies, or policymakers) can share their problems and possible solutions. A clear set of questions, related to shared problems, often enhances communication and the creation of trust. This process of creating trust among the participants often takes a substantial period of time.[21]

Finally, the interview results highlighted the need for flexibility as a procedural feature of international ERD3 cooperation activities. International ERD3 cooperation projects are evolutionary processes, and it is not uncommon for unforeseen developments to occur and dramatically change the impact and direction of the innovation process. A procedure that creates trust and commitment among participants can more easily cope with these kinds of changes.

5.5.4. Evaluating the Impacts of ERD3 Cooperation

Once a decision has been made to launch a particular international ERD3 cooperation project – which may be made using a broad range of decision criteria, as discussed

[21] One interviewee mentioned that it took five years until all participants really trusted each other, but the creation of this shared belief was a major factor in the successful outcome of a large international cooperation project.

Table 5.6. *Evaluation criteria used by United Nations Development Programme and the International Energy Agency*

Project evaluation	Outcome evaluation	Technology progress
• Strategic contribution • Substantiveness and comprehensiveness • Justification of transaction costs • Fulfilment of contractual and management obligations • Objectives, strategic plan, and work programs in place	• Relevance of strategy and work plan to market • Accelerating technology deployment • Industry participation • Dissemination of results • Reduced or avoided research costs • Positive return on investments • Improved cost reduction • Unique results • Improved interaction • Improved access to personnel, training, information, technology, or equipment	• Contribution to technology evolution • Citation and awards • Use of results by institutions • Significant success stories • Spillovers to other energy technology areas

previously – what criteria can be used to evaluate the project's progress? Here, too, the wide range of objectives and types of projects makes it problematic to attempt to focus on a single, unified design and evaluation approach.

In Table 5.6, we have classified IEA criteria for evaluating their Implementing Agreements into three categories: (1) project evaluation; (2) outcome evaluation; and (3) technology progress criteria. The first two categories are common evaluation categories for international development aid projects (UNDP, 2009), and they are also present within IEA's "structural and management criteria" and IEA's "value-based criteria." The latter category is specific to the IEA and attempts to measure progress in projects that include basic science and research.

There is little systematic evaluation of international cooperation projects by the U.S. government or by most other governments. For example, there are no requirements to evaluate the outcomes of formal international science and technology agreements signed by the State Department or created during participation in international fora (State, 2001). Many activities have internal evaluation reports, though these are generally not publicly available and may use quite different evaluation approaches from one activity to the next. Other international cooperation activities are also evaluated on a yearly or biannual basis (e.g., the U.S.–Brazil MOU to Advance Cooperation on Biofuels), but these evaluations are not systematic and cannot be used to compare projects to each other, or to learn lessons about what works and what does not, under what circumstances, in carrying out such projects.

There are several reasons why evaluation of international ERD3 cooperation projects is difficult. First, as noted earlier, many international cooperation activities supported by the U.S. government do not involve the development of new technologies. The benefits of projects focused on smoothing the path for increased deployment of existing technologies are difficult to evaluate, as even if increased deployment occurs, the project may not have been an important cause of that result. Even where projects *are* focused on developing new technologies, it is difficult to assess how much impact their achievements have on subsequent improvements of cost and performance of technologies available in the market.

Second, collecting and analyzing information about activities and results takes effort. Even if participants are interested in disseminating their work and allowing other people to evaluate it, the process of collecting, updating, and coordinating information is cumbersome. In those cases in which project evaluation is mandatory, the information that is currently being collected is not necessarily useful for other stakeholders or for future cooperation projects. For example, the IEA Implementing Agreements are reviewed every five years, and their most significant achievements are published every two years. However, the overview reports of success stories provide little systematic analysis of progress made, the problems tackled, the solutions provided, lessons learned about the best approaches to such cooperation, or how projects are related to other international cooperation activities. Instead, the reports provide incremental updates of previous project descriptions or a summation of outputs without contextualizing applicability.[22] Similarly, foreign or international initiatives to create "information portals" lack the sophistication that is required to aid policymakers at different levels throughout the government or innovators in their decisions to design or engage in international ERD3 cooperation activities.[23]

Third, significant portions of the benefits of many international ERD3 cooperation activities are intangible. The value of a cooperation project is often the relationship itself, and it is important to also assess the impact of a project on networks and relationships. One interviewee mentioned that the network established to set up an international cooperation project was as important, or even more important, than the project itself. Because of this network, the interviewee assisted high-level policymakers in Brazil in the development of integrated resource planning, relying on both

[22] It is very important to note that "collecting information" does not equate to an evaluation. A bad example of an evaluation effort is the Interagency Working Group (IAWG), which collects information annually on all international exchange and training programs within the U.S. government. These reports are very long (200 to 450 pages), have little structure, and do not explicitly provide lessons and insights. In other words, they create a lot of work but add little value.

[23] Several IEA Implementing Agreements (like the greenhouse gas R&D program or the Clean Coal Centre) have built online databases to facilitate international ERD3 cooperation, but these databases are structured around technologies or "types of information" (publications, policy documents, meetings, etc.) and do not break down the data into tangible information for policymakers or participants. Similarly, EC's initiatives to create databases of R&D policies (ERAWATCH) or the Strategic Energy Technology Information System (SETIS) are structured around broad technology areas or countries rather than specific information needs.

hydropower and biomass sources for electricity. Without any international cooperation projects, the interviewee would not have had the connections and standing to assist and to influence the decisions of the Brazilian officials.

Despite these difficulties, interviewees acknowledged the need for reporting and evaluating. Indeed, an analysis of several international cooperation activities focused on sustainable development issues concluded that institutionalizing mechanisms for communication, translation, and mediation has proven to be crucial for enhancing the effectiveness of international innovation activities (Cash et al., 2003). Some interviewees argued that an annual timeframe for reporting allows participants sufficient time to make progress, to contain any administrative burden, to refocus their activities, and to reflect on their progress. Finally, as one interviewee pointed out, any evaluation of project management needs to take into consideration the fact that the success rate of any innovation process is unlikely to be high, and that international innovation partnerships have an even lesser chance of succeeding. In such uncertain situations, assessment and evaluation are the only means by which investigators can learn from and reflect on the value of these high-risk projects, even if they do not achieve their full initial goals or ambitions.

These findings suggest that the focus on evaluation has to shift away from measuring generic project, outcome, and technology progress metrics to assessing indicators of outcomes that are specific to the participants' goals and objectives. These outcome indicators should allow participants to assess the value of these projects, to reflect on the direction of the project as well as on their own contributions and expectations, and they should be formulated so that nonparticipants can understand and learn from them as well. Finally, project evaluation criteria should support the managers of the projects while they are active in the project, and should provide feedback to government officials after the projects have finished.

5.6. A Strengthened Approach for the U.S. Government

In the complex and dynamic global ERD3 environment just described, the U.S. government must seek to accomplish three goals: (1) finding the right balance between competition and cooperation; (2) identifying as many of the most promising opportunities for cooperation to accelerate innovation as possible; and (3) coordinating and prioritizing the effort, targeting substantial resources to the most important strategic priorities. In this section, we offer recommendations in each of these areas.

5.6.1. Balancing Competition and Cooperation

As noted at the outset of this chapter, the opportunities from international ERD3 cooperation are huge – joining ideas from multiple sources, pooling resources, taking advantage of the best locations for demonstration facilities, and more. But the competitive pressures are great as well, with the major players in energy technology each jockeying for their share of the energy technology markets of the 21st century. Choices about where to compete and where to cooperate will have to be made case by case, using the best judgment of the officials involved. Here, we offer only a brief discussion of some of the key issues to be considered.

Some of the most challenging issues relate to whether the U.S. government should promote technology transfer, so developing countries can manufacture clean energy technologies themselves, or focus on protecting U.S. technological advantages. As a study from the Council on Foreign Relations pointed out, neither an all-transfer nor an all-protection strategy is likely to be effective or sustainable (Levi, Economy, O'Neil, & Segal, 2010). Instead, they recommend an approach that combines efforts to convince emerging economies to open their markets to technologies and investments from U.S. firms and protect IPR – addressing the commercial concerns of U.S. firms and workers – while at the same time actively promoting technology transfer where appropriate, offering benefits to firms in the emerging economies. But where states restrict their markets to the disadvantage of U.S. firms, and provide little protection for intellectual property, transferring technology poses bigger risks and fewer advantages.

China poses some of the most difficult conundrums for U.S. policy in balancing competition and cooperation. It is the world's largest and fastest-growing consumer of energy – a crucial market for energy technology firms. It is making huge investments in energy technology innovation, and the opportunities for cooperation with China's scientists and engineers across a wide range of energy technologies are immense. At the same time, the Chinese government has aggressively supported its firms in building up their competitive position – buying technologies abroad, providing extensive support to domestic firms to strengthen their competitive position, demanding that Western firms transfer technology as a condition for increased access to the Chinese market, and more. (Some of these approaches are discussed in Levi, Economy, O'Neil, & Segal, 2010.) Since 2000, China has gone from a trivial producer of solar panels to manufacturing two-thirds of the world's supply; seized an important fraction of the market for wind turbines; become an important player in the market for electric vehicles; acquired the technologies that will soon allow it to become a competitive exporter of nuclear power plants; and more. As noted earlier, governments and firms in the United States and Europe have increasingly been crying foul, arguing that some of China's support for its firms may violate WTO rules.

There are clearly areas where the United States needs to push back on such behavior from China and other countries, seeking to ensure that all participants play by fair international rules. But international cooperation itself can provide a powerful tool for addressing some of these issues. Personal relationships and exposure to another party's thinking and concerns developed through cooperation can lead to policy change. In the area of IPR protection, for example – long one of the most hotly contested technology issues between the United States and China – the U.S.–China Clean Energy Research Center (CERC) has sponsored a number of workshops and developed a significant degree of common understanding of the importance of strong IPR protection to the incentives for innovation (CERC, 2013). Chinese incentives are also changing: as Chinese innovators and firms generate more and more of their own intellectual property and depend less on taking advantage of foreign IPR, there is more internal demand for effective IPR protection.

Similarly, at the commercial level, through the interactions that arise from cooperation, many issues have been successfully addressed. Companies in both the United States and Europe have found that they can increase, rather than decrease, their commercial advantages by matching their strengths to those of partners in China and elsewhere, just as Apple has chosen Chinese manufacturers to produce its iPhones. The U.S. government should seek to limit the degree to which other governments put undue pressure on such commercial deals (as in the cases of China and other countries limiting access to their markets unless foreign firms agree to transfer major parts of their technology), but it would be pointless and counterproductive to try to stop such global integration of the clean energy value chain from occurring.

Several considerations can help guide U.S. policymakers as they seek to find the best path forward, with China and with other countries. The key issues vary depending on the stage of innovation a particular project involves, and what the United States and its potential partners or competitors are bringing to the table.

In the case of *early stage R&D*, well before technologies become commercial, international R&D cooperation brings many advantages and carries relatively few risks of undermining U.S. technological advantages. Such projects pool ideas to generate new knowledge that benefits all the participating partners (and frequently spills over to others as well). It is important for the parties to agree on how the intellectual property generated in such cooperative projects will be handled, but at this stage that is often not unduly difficult.

At the *demonstration* phase, the balance between competition and cooperation is conceptually more difficult, as technologies are nearing the point of commercial application, and questions of who will get the commercial benefits from them inevitably arise. In practice, however, the U.S. government has few funds to pay for large-scale technology demonstrations in the coming years, and there may be great advantages in sharing the costs of such efforts with other countries. China, in particular, has shown an ability to mobilize resources for such projects – and in some cases it may be possible to implement some projects more cheaply there. From the perspective of U.S. interests, a demonstration of a technology developed in the United States that is carried out in China – with careful negotiation of which parties will receive what share of the resulting intellectual property – is clearly preferable to the technology remaining only at the stage of computer analyses and lab-scale experiments. In expensive projects of this kind, working out who will make what investments and who will own what share of the resulting intellectual property is a natural and inevitable part of structuring any project involving multiple players.

At the stage of development and deployment of commercial energy systems, markets should largely drive decisions on whether and how to share technology internationally. Where other governments intervene in market decisions to support their firms to the disadvantage of U.S. firms, it makes sense for the U.S. government to seek ways to push back without unduly undermining the flow of innovation. The U.S. government can also help play matchmaker when cooperation appears to make sense, and use available tools to help support U.S. exports. But overall, market actors should play the leading role. Indeed, it is important for U.S. policymakers to recognize the limits of the U.S. government's role.

What factors, then, should U.S. government decision makers consider in making such decisions? Here are a few:

- How much opportunity for accelerating development of new technologies does the proposed cooperation offer?
- How much could the proposed cooperation lead to increased access to important foreign markets?
- What U.S. technological advantages might be risked by a particular transfer or cooperative project, how might those risks be mitigated, and how important would the resulting change in relative market opportunities be?
- How committed is the country with whom the cooperation would take place to genuine cooperation in the common interest – as opposed to a zero-sum search for competitive advantage? In particular, how open are their markets to sales and investments by U.S. firms in the area in question, how much have they used market power to force U.S. firms to make technological concessions, and how committed is the government to genuinely effective protection of IPR?

Policymakers will have to draw the balance between cooperation and competition case by case, adapting their approach on the basis of experience. But a focused approach to answering questions such as these can help in making these choices.

5.6.2. A Bottom-up Approach

New opportunities for international ERD3 cooperation arise at many different levels within the U.S. government, as well as in many different places around the world. High-level policymakers cannot be expected to be aware of every potentially important opportunity. Only a bottom-up approach that harnesses the insights of many different U.S. innovators involved in international cooperation networks is flexible enough to seize these opportunities, and, as such, to maximize the benefits the United States can gain from international ERD3 cooperation.

But a bottom-up approach provides little opportunity to prioritize; minimize overlaps; find and fix gaps; or learn from experience about what works, what does not, and under what circumstances. A top-down approach, based on information about bottom-up activities and designed as a complement, rather than an alternative, could address some of the weaknesses of an exclusively bottom-up approach. Hence, we suggest a two-track approach, incorporating both bottom-up and top-down elements. The aim of this approach is to support bottom-up activities, collect bottom-up information, and use this information to guide interagency coordination mechanisms that collect, assess, and evaluate U.S. government involvement in international ERD3 cooperation. The interagency approach we propose would also have the ability to instigate international ERD3 cooperation efforts that fill gaps or seize synergies among existing national and international ERD3 activities.

To accelerate the bottom-up identification of new opportunities – and to provide the funds necessary for the initial explorations and building of personal relationships that are often needed to get a new initiative started – the U.S. government should direct each department or agency that is funding ERD3 activities to create an "incubator fund," setting aside a small portion of its energy technology innovation budget to support international projects, based on opportunities identified by working-level project managers and experts. This approach would significantly increase the total funding directed to international cooperation efforts.

Whenever an expert involved in ERD3 projects had an idea for a new cooperative project, his or her office could apply to the incubator fund for the support needed to get it off the ground. Because each department has different objectives, each department should control its own fund – but the funds should share their practices, lessons learned, and administrative procedures. The structure for administration should be designed to enable quick decisions about allocating funds.

Even beyond these incubator funds, departments and agencies should always be on the lookout for situations where international ERD3 cooperation could advance their energy innovation objectives. In general, these activities include:

- Energy innovation activities that require specific local knowledge
- Energy innovation activities and expertise of interest to U.S. innovators that are in foreign countries
- The design or operation of expensive test facilities
- High-cost energy RD&D activities, such as demonstration projects
- Activities that support the early deployment of energy technologies in new markets

At the same time, the U.S. government should give particular labs or firms the job of scanning the environment in particular areas of energy technology to: identify firms and institutes that are doing especially promising work; assess technological trends and developments that may open new opportunities or make other approaches currently under development obsolete (or at least less competitive); and find potential opportunities for partnerships. Venture capital firms do this kind of scanning of the technological horizon routinely, to identify investment opportunities and assess the competition to the firms they are backing. U.S. labs and firms are already doing some of this, but it does not appear to be integrated into central repositories of information focused on supporting international ERD3 cooperation.

5.6.3. A Top-down Approach

A decentralized, bottom-up approach can involve many innovators in identifying opportunities for international cooperation. But by itself, such an approach could lead to duplication in some areas and gaps in others, and may not be able to mobilize sufficient resources behind key strategic priorities. We therefore recommend that the U.S. government establish a top-down approach by creating an interagency task force to (1) identify and mobilize resources behind key strategic priorities in international ERD3 cooperation, and ensure that these are managed effectively and sustained over time; and (2) to the extent practicable, coordinate international ERD3 cooperation to minimize duplication and seize opportunities for synergies, both within these efforts and between the international efforts and purely domestic U.S. ERD3 programs. The task force should work directly with the Office of Management and Budget (OMB) to ensure that the initiatives it identifies as the highest priorities will receive adequate funding. It should be responsible for identifying and communicating strategic guidelines for cooperation on energy technology innovation every year and for producing an annual a report to Congress, the White House, the different departments, and the public.

One approach to establishing such a task force would be to locate it within the National Science and Technology Council (NSTC) managed by the Office of Science and Technology Policy (OSTP).[24] The task force should be co-chaired by DOE and OSTP, the department and the White House office with the most extensive involvement in international ERD3 activities. The task force should have active participation from the other agencies involved, including the State Department, the U.S. Department of Commerce, the U.S. Department of Agriculture, and the U.S. Agency for International Development, as well as private sector actors.

5.6.4. Connecting the Bottom and the Top

To be most effective, it is important for the actors at both the bottom and the top of the decision-making structure to have information on the different activities underway and lessons learned from implementing them.

Hence, we recommend that the U.S. government establish a new information-sharing platform to collect and disseminate information on international ERD3 activities, in order to: (1) ensure continuity and follow-up between activities; (2) avoid

[24] For a similar recommendation, see PCAST 2010.

duplication and repetition of activities supported across the government; (3) create complementarities between and across projects, programs, and agreements; and (4) provide opportunities for analysis and sharing of lessons learned, so that implementation of these efforts can learn and adapt over time. Since multiple U.S. agencies are active at each level, it is necessary to provide an information system that crosses agency borders and includes information from high-level officials, program officers, and project managers in all U.S. agencies. A management system to collect and distribute information about existing and new agreements, programs, and projects should provide comparable and coherent information, while at the same time acknowledging that information needs are different at each level.

There are several options for this information platform's institutional home; one option would be the Energy Information Administration, which is already in the business of collecting and analyzing large quantities of energy-related data (though it does not yet have experience in assessing lessons learned and best practices in direct support of policymaking). The system should collect information about existing international cooperation projects, programs, and high-level agreements across the different U.S. agencies.

Incentives should be put in place to ensure that those responsible for the projects consistently provide information on the details of the projects and their progress. The platform should also enable private firms and not-for-profit organizations to access and share information about their projects. Information needs to be comparable at different levels and across agencies, and accessible to both U.S. policymakers and other stakeholders who may be interested in cooperating with other parties internationally.

In addition to collecting and making this information available, the U.S. government should follow four general principles in seeking to manage and integrate these bottom-up and top-down international ERD3 cooperation activities. First, despite the diversity of objectives and types of projects underway, an effective strategy should attempt to create *coherence* between the different government departments and agencies, and in particular, as just noted, to collect, analyze, and disseminate coherent information about them. Second, the U.S. government should attempt to create *complementarities* between the activities taking place at the different governance levels (agreements, programs, and projects). Third, the U.S. government should attempt to create *synergies* among the international activities it supports and between them and its domestic ERD3 efforts. Fourth, the U.S. government should seek to integrate the initiatives started from the bottom up and the priorities established from the top down; each new bottom-up effort should connect to existing innovation activities by either supporting existing collaborations or addressing existing gaps. Together, these four guiding principles, underpinned by their supporting mechanisms, can create an environment in which U.S. policymakers and U.S. innovators work together to capture the benefits and mitigate

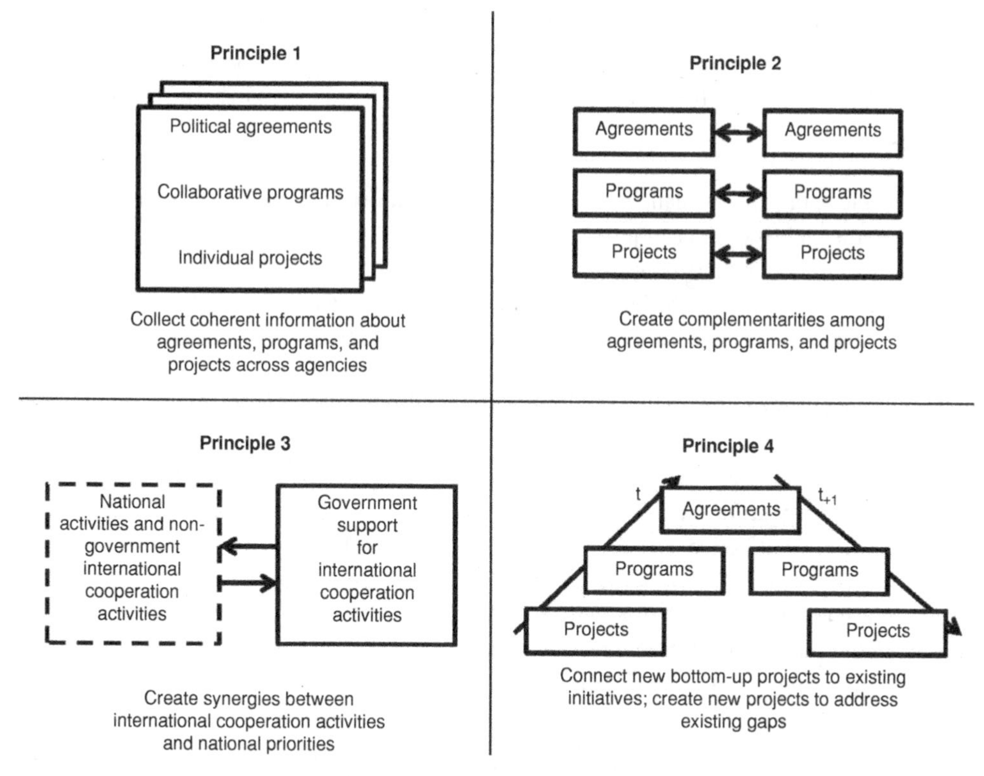

Figure 5.3. Four guiding principles for managing international cooperation activities within the U.S. government.

the risks of international ERD3 cooperation to address U.S. energy challenges in an increasingly complex and interdependent world. Figure 5.3 provides a visual representation of these four principles.

5.7. Conclusions and Recommendations

The United States needs a more adaptive and effective approach to international ERD3 cooperation to meet the challenges of energy innovation in a rapidly changing international landscape. The discussion in this chapter has made it clear that greater coordination and learning from experience is needed – but also that there are so many different goals of international ERD3 cooperation, and so many actors engaged in it, that it is unlikely to be practical to centralize all decisions on what efforts to undertake.

Instead, we recommend that the U.S. government adopt a two-track approach that supports international ERD3 activities by U.S. innovators from the bottom up, and simultaneously supports U.S. policymakers by addressing potential gaps or synergies from the top down. Both approaches require a more diligent and systematic effort of collecting and disseminating information about what projects are currently supported, what problems they address, and what solutions they may provide. In particular, we propose:

- A strategic approach to balancing cooperation and competition, asking key questions about the benefits and risks of each project, while understanding that cooperation itself can be a powerful tool to convince states to modify some uncompetitive practices
- Setting aside a portion of the budgets of each major energy technology innovation program for international cooperation to finance efforts identified bottom-up by experts involved in international energy RD&D efforts
- A new interagency task force to establish priorities, minimize overlaps, seize opportunities for synergy, and lead the overall effort
- A new system to collect information on international ERD3 cooperation activities that will increase participants' awareness of the work underway and increase the chance for programs to become more effective over time by learning from experience.

Together, the recommendations in this chapter will help the U.S. approach to international ERD3 cooperation meet the CASCADES set of criteria for an effective energy innovation strategy outlined in Chapter 1. By helping to identify and fill gaps, they will help ensure that U.S. efforts are *comprehensive* enough not to miss key opportunities. By enabling the information collection and analysis needed to learn by doing, these steps will help ensure that U.S. efforts are *adaptive* to experience and to new conditions. By helping to identify strategic priorities and align them with resources, they should help ensure that these efforts are *sustainable.* Though none of the recommendations in this chapter are explicitly focused on making U.S. efforts more *cost-effective*, both the bottom-up identification of key opportunities and the focused collection and analysis of information on how programs are working should help in achieving that objective. The creation of a high-level task force that can scan the environment and change course when a particular effort fails or a new opportunity arises should make these efforts more *agile.* In the same way that our recommendations ensure that U.S. efforts take a comprehensive approach to different actors and stages of innovation, they should also help ensure that the portfolio is *diversified* in its coverage of different technologies, and *equitable* between one technology or one set of stakeholders and another – particularly because the information collection and analysis will contribute to making clear judgments about the performance of one set of efforts versus another. Finally, the creation of the high-level task force to set strategic priorities should certainly make these efforts more *strategic* and focused on the greatest opportunities for potential gain. Overall, it is our hope that the recommendations in this chapter will help U.S. policymakers use international ERD3 cooperation as a powerful tool to accelerate the development and deployment of new and improved energy technologies in the United States and around the world.

References

AEIC. (2010). "A business plan for America's energy future." Washington, D.C.: American Energy Innovation Council. Retrieved from http://www.americanenergyinnovation.org/wp-content/uploads/2012/04/AEIC_The_Business_Plan_2010.pdf (Accessed March 1, 2014).

Anadon, L.D. (2012). "Missions-oriented RD&D institutions in energy: A comparative analysis of China, the United Kingdom, and the United States." *Research Policy*, 41(10):1742–56.

Anadon, L.D., Bunn, M., Chan, G., Chan, M., Jones, C., Kempener, R., Lee, A., Logar, N., & Narayanamurti, V. (2011, November). *Transforming U.S. Energy Innovation*. Cambridge, Mass.: Energy Technology Innovation Policy, Belfer Center for Science and International Affairs, John F. Kennedy School of Government, Harvard University.

Archibugi, D., & Iammarino, S. (1999). "The policy implications of the globalisation of innovation." *Research Policy*, 28(2–3):317–36.

Archibugi, D., & Iammarino, S. (2002). "The globalization of technological innovation: Definition and evidence." *Review of International Political Economy*, 9(1):98–122.

Asia Society. (2009). "A roadmap for U.S.-China cooperation on energy and climate change." Arlington, Va.: Pew Center on Global Climate Change. Retrieved from http://www.c2es.org/docUploads/US-China-Roadmap-Feb09.pdf (Accessed March 1, 2014).

Atkinson, M., Castellas, P., & Curnow, P. (2009). "Independent review of Asia-Pacific partnership flagship projects." Prepared for the Department of Resources, Energy and Tourism on behalf of the Asia-Pacific Partnership on Clean Development and Climate (APP). Retrieved from http://www.asiapacificpartnership.org/pdf/resources/Final_Flagship_Review_-_English.pdf (Accessed March 1, 2014).

Atkinson, R., Shellenberger, M., Nordhaus, T., Swezey, D., Norris, T., Jenkins, J., Ewbank, L., Peace, J., & Borofsky, Y. (2009). "Rising tigers, sleeping giant." Oakland, Calif.: The Breakthrough Institute. Retrieved from http://thebreakthrough.org/blog/Rising_Tigers.pdf (Accessed March 1, 2014).

Bazilian, M., De Coninck, H., Cosbey, A., & Neuhoff, K. (2009). *Mechanisms for International Low Carbon Technology Cooperation: Roles and Impacts*. London: Climate Strategies. Retrieved from http://www.climatestrategies.org/research/our-reports/category/43/221.html (Accessed March 1, 2014).

Beamish, P.W., & Killing, J.P. (1996). "Introduction to the special issue." *Journal of International Business Studies*, 27(5):iv–xxxi.

Beinhocker, E., Oppenheim, J., Irons, B., Lahti, M., Farrell, D., Nyquist, S., Remes, J., Naucler, T., & Enkvist, P. (2008, June). "The carbon productivity challenge: Curbing climate change and sustaining economic growth." McKinsey Global Institute, McKinsey Climate Change Special Initiative. Retrieved from http://www.mckinsey.com/insights/energy_resources_materials/the_carbon_productivity_challenge (Accessed March 1, 2014).

Belderbos, R., Kyoki, F., & Iwasa, T. (2009). "Foreign and domestic R&D investment." *Economics of Innovation & New Technology*, 18(4):369–80.

Betsill, M.M. (2010). "Introduction: Is the Asia-Pacific partnership a viable alternative to Kyoto?" An editorial essay. *Wiley Interdisciplinary Reviews: Climate Change*, 1(1):9–9.

Biermann, F., Pattberg, P., van Asselt, H., & Zelli, F. (2009). "The fragmentation of global governance architectures: A framework for analysis." *Global Environmental Politics*, 9(4):14–40.

BP. (2011). "Statistical review of world energy, 2011." Retrieved from http://www.bp.com/assets/bp_internet/globalbp/globalbp_uk_english/reports_and_publications/statistical_energy_review_2011/STAGING/local_assets/pdf/statistical_review_of_world_energy_full_report_2011.pdf (Accessed March 1, 2014).

Cash, D.W., Clark, W.C., Alcock, F., Dickson, N.M., Eckley, N., Guston, D.H., Jäger, J., & Mitchell, R.B. (2003). "Knowledge systems for sustainable development." *Proceedings of the National Academy of Sciences of the United States of America*, 100(14):8086–91.

CCTP. (2006). *U.S. Climate Change Technology Program Strategic Plan*. DOE/PI-0005. Washington, D.C.: U.S. Climate Change Technology Program. Retrieved from http://www.climatetechnology.gov/stratplan/final/CCTP-StratPlan-Sep-2006.pdf (Accessed March 1, 2014).

CCTP. (2009). *Strategies for the Commercialization and Deployment of Greenhouse Gas Intensity-Reducing Technologies and Practices*. Committee on Climate Change Science and Technology, Department of Energy. Retrieved from http://energy.gov/sites/prod/files/CDStratCompleteReport11609%5B2%5D.pdf (Accessed March 3, 2014).

CEPP. (2009). "Who owns the clean tech revolution?" Center for Environmental Public Policy, Goldman School of Public Policy, University of California, Berkeley. Retrieved from http://cal.gspp.berkeley.edu/IPR/whoowns.pdf (Accessed March 3, 2014).

CERC. (2013). *Annual Report 2012–2013*. U.S.-China Clean Energy Research Center. Retrieved from http://www.energetics.com/news/Documents/US-China_CERC_Annual_Report_2012-2013.pdf (Accessed March 3, 2014).

Chikkatur, A.P., & Sagar, A.D. (2007). *Cleaner Power in India: Towards a Clean-Coal-Technology Roadmap*. Energy Technology Innovation Policy Group, Belfer Center for Science and International Affairs, John F. Kennedy School of Government, Harvard University. Retrieved from http://belfercenter.ksg.harvard.edu/files/Chikkatur_Sagar_India_Coal_Roadmap.pdf (Accessed March 3, 2014).

Chikkatur, A.P., Sagar, A.D., & Sankar, T.L. (2009). "Sustainable development of the Indian coal sector." *Energy*, 34(8):942–53.

Chum, H. (2009). "Presentation spotlight: U.S.-Brazil MOU to Advance Cooperation on Biofuels." In Global Innovation Initiative, *Catalyze: U.S.-Brazil Innovation Learning Laboratories*, 95. Golden, Col.: National Renewable Energy Laboratory. Retrieved from http://www.compete.org/images/uploads/File/PDF%20Files/Catalyze._U_.S_.-Brazil_Innovation_Learning_Laboratories_.pdf (Accessed March 3, 2014).

Conacyt. (2008). *Informe Genereal de Estado de la Ciencia y la Tecnologia*. Gobierno Federal, Mexico: Consenjo National de Ciencia y Tecnologia.

Das, T.K., & Teng, B.S. (1998). "Between trust and control: Developing confidence in partner cooperation in alliances." *The Academy of Management Review*, 23(3):491–512.

David, A.S. (2009, June). *Wind Turbines: Industry and Trade Summary*. Pub. ITS-02. Washington, D.C.: U.S. International Trade Commission. Retrieved from http://www.usitc.gov/publications/332/ITS-2.pdf (Accessed March 1, 2014).

Department of Public Enterprises. (2010). Address by the Minister of Public Enterprises, Barbara Hogan, to the National Assembly, on the Pebble Bed Modular Reactor. September 16, 2010. Retrieved from http://www.dpe.gov.za/news-971 (Accessed March 3, 2014).

DOC. (2010a). *Annual Report on Technology Transfer: Approach and Plans, Fiscal Year 2009 Activities and Achievements*. U.S. Department of Commerce. Retrieved from http://www.nist.gov/tpo/publications/upload/2009-Tech-Transfer-Rept-FINAL.pdf (Accessed March 3, 2014).

DOC. (2010b). Charter of the renewable energy and energy efficiency Advisory Committee. International Trade Agency, Department of Commerce. Retrieved from http://export.gov/static/REEEAC%20II%20CharterFinal%20Signed_Latest_eg_main_050884.pdf (Accessed March 3, 2014).

DOE. (2011). *U.S.-China Clean Energy Cooperation: A Progress Report by the U.S. Department of Energy*. U.S. Department of Energy. Retrieved from http://www.us-china-cerc.org/pdfs/US_China_Clean_Energy_Progress_Report.pdf (Accessed March 3, 2014).

Edquist, C. (1997). *Systems of Innovation: Technologies, Institutions, and Organizations*. New York: Pinter.

EGTT. (2009). "Strategy paper for the long-term perspective beyond 2012, including sectoral approaches, to facilitate the development, deployment, diffusion and transfer of technologies under the Convention." Report by the Chair of the Expert Group on Technology Transfer. Retrieved from http://unfccc.int/resource/docs/2009/sb/eng/03.pdf (Accessed March 3, 2014).

Gallagher, K.S. (2014). *The Globalization of Clean Energy Technologies: Lessons From China*. Cambridge, Mass.: MIT Press.

Gallagher, K.S., Anadon, L.D., Kempener, R., & Wilson, C. (2011). "Trends in investments in global energy research, development, and demonstration." *Wiley Interdisciplinary Reviews: Climate Change*, 2(3):373–96.

Gallagher, K.S., Gruebler, A., Kuhl, L., Nemet, G., & Wilson, C. (2012). "The energy technology innovation system." *Annual Review of Environment and Resources*, 37:137–62.

Gallagher, K.S., Holdren, J.P., & Sagar, A.D. (2006). "Energy-technology innovation." *Annual Review of Environment and Resources*, 31(1):193–237.

Gopalakrishan, A. (2000, March). "For synergy in energy." *Frontline*, 17(6):18–31.

Haas, E.B. (1975). "Is there a hole in the whole? Knowledge, technology, interdependence and the construction of international regimes." *International Organization*, 29(3): 827.

Hagedoorn, J. (2006). "Understanding the cross-level embeddedness of interfirm partnership formation." *Academy of Management Review*, 31(3):670–80.

Hekkert, M.P., Suurs, R.A.A., Negro, S.O., Kuhlmann, S., & Smits, R.E.H.M. (2007). "Functions of innovation systems: A new approach for analysing technological change." *Technological Forecasting and Social Change*, 74(4):413–32.

Huang, Z. (2011). "Clean energy subsidies and the law of the WTO." Presentation, March 31. Cambridge, Mass.: Harvard China Project, Harvard School of Engineering and Applied Sciences.

IEA. (2005). *International Energy Technology Collaboration and Climate Change Mitigation: Synthesis Report*. International Energy Agency. Organisation for Economic Co-operation and Development. Retrieved from http://www.oecd.org/dataoecd/62/42/35798801.pdf (Accessed March 3, 2014).

IEA. (2010a). *Energy Technology Initiatives: Implementation through Multilateral Co-operation*. International Energy Agency. Retrieved from http://www.iea.org/publications/freepublications/publication/technology_initiatives-1.pdf (Accessed March 3, 2014).

IEA. (2010b). IEA Addressing Climate Change Database. Retrieved from http://www.iea.org/policiesandmeasures/climatechange/ (Accessed March 3, 2014).

IEA. (2010c). IEA Clean Coal Projects Database. Retrieved from http://www.iea-coal.org.uk/site/ieacoal/databases/clean-coal-projects (Accessed March 3, 2014).

IEA. (2010d). IEA Energy Efficiency Database. Retrieved from http://www.iea.org/policiesandmeasures/energyefficiency/ (Accessed March 3, 2014).

IEA. (2010e). IEA Global Renewable Energy Database. Retrieved from http://www.iea.org/policiesandmeasures/renewableenergy/ (Accessed March 3, 2014).

IEA. (2010f). *World Energy Outlook 2010*. Paris: IEA, OECD. Retrieved from http://www.worldenergyoutlook.org/ (Accessed March 3, 2014).

IEA. (2011). *Energy Technology RD&D Budgets* (2011 edition). Detailed Country RD&D Budgets. Retrieved from http://www.iea.org/stats/rd.asp (Accessed March 3, 2014).

IEA. (2012). *World Energy Outlook 2012*. Paris: IEA, OECD. Retrieved from http://www.worldenergyoutlook.org/ (Accessed March 3, 2014).

Iida, K. (1993). "Analytic uncertainty and international cooperation: Theory and application to international economic policy coordination." *International Studies Quarterly*, 37(4): 431–57.

Interagency Task Force on Carbon Capture and Storage. (2010). *Report of the Interagency Task Force on Carbon Capture and Storage*. U.S. Department of Energy. U.S. Environmental Protection Agency. Retrieved from http://www.fe.doe.gov/programs/sequestration/ccstf/ CCSTaskForceReport2010.pdf (Accessed March 3, 2014).

IWG. (2010). "Growing America's fuel." Biofuels Interagency Working Group. U.S. Department of Agriculture, U.S. Department of Energy, U.S. Environmental Protection Agency. Retrieved from http://www.whitehouse.gov/sites/default/files/rss_viewer/growing_americas_fuels.PDF (Accessed March 3, 2014).

Johnson, A., & Jacobsson, S. (2001). "Inducement and blocking mechanisms in the development of a new industry: The case of renewable energy technology in Sweden." In *Technology and the Market: Demand, Users and Innovation*, edited by R. Coombs, pp. 89–111. Cheltenham, UK: Edward Elgar.

Kasprzycki, D., Ozegalska-Trybalska, J., & Mayr, A. (2008). "How proactive intellectual property management can improve research collaborations: Good practices in EU and BRIC higher educational institutions." IP-Unilink. Retrieved from http://www.ip-unilink.net/public_documents/Good_Practice_Guide_web.pdf (Accessed March 3, 2014).

Kempener, R., Anadon, L.D., & Condor Tarco, J. (2010). "Governmental energy innovation investments and policies in the major emerging economies: Brazil, Russia, India, Mexico, China, and South Africa." Energy Research, Development, Demonstration and Deployment Project, Energy Technology Innovation Policy Group, Belfer Center for Science and International Affairs. Retrieved from http://belfercenter.ksg.harvard.edu/publication/20517/ (Accessed March 3, 2014).

Levi, M.A., Economy, E.C., O'Neil, S.K., & Segal, A. (2010). "Energy innovation: Driving technology competition and cooperation among the U.S., China, India and Brazil." Council on Foreign Relations. Retrieved from http://www.cfr.org/publication/23321/energy_innovation.html (Accessed March 3, 2014).

Markusson, N., Issii, A., & Stephens, J.C. (2011). "The social and political complexities of learning in CCS demonstration projects." *Global Environmental Change*, 21:293–302.

Metz, B., Davidson, O., Martens, J., Van Rooijen, S., & Van Wie Mcgrory, L. (Eds.) (2000). *Methodological and Technological Issues in Technology Transfer*. Cambridge: Cambridge University Press.

Nelson, R.R., & Chandler, A. (1993). *National Innovation Systems: A Comparative Analysis*. New York: Oxford University Press.

NRC. (1999). "The pervasive role of science, technology, and health in foreign policy: Imperatives for the Department of State." Committee on Science, Technology, and Health Aspects of the Foreign Policy Agenda of the United States, National Research Council. Retrieved from http://www.nap.edu/catalog/9688.html (Accessed March 3, 2014).

NREL. (2008). "Strengthening U.S. leadership of international clean energy cooperation." National Renewable Energy Laboratory. Retrieved from http://www.nrel.gov/international/pdfs/44261.pdf (Accessed March 3, 2014).

NSB. (2001, November 15). "Towards a more effective role for the U.S. government in international science and engineering." National Science Board. Retrieved from http://www.nsf.gov/nsb/documents/2002/nsb01187/pdfs/nsb01187.pdf (Accessed March 3, 2014).

NSB. (2008, February 14). "International science and engineering partnerships: A priority for U.S. foreign policy and our nation's innovation enterprise." National Science Board. Retrieved from http://www.nsf.gov/nsb/publications/2008/nsb084.pdf (Accessed March 3, 2014).

NSF. (2008, December). "Research and development in industry: 2004." Division of Science Resources Statistics (SRS), National Science Foundation. Retrieved from http://www.nsf.gov/statistics/nsf09301/ (Accessed March 3, 2014).

Ockwell, D.G., Haum, R., Mallett, A., & Watson, J. (2010). "Intellectual property rights and low carbon technology transfer: Conflicting discourses of diffusion and development." *Global Environmental Change*, 20(4):729–38.

OECD. (2010). "Aggregate official and private flows." Organisation for Economic Cooperation and Development. Retrieved from http://stats.oecd.org/Index.aspx?DataSetCode=CRSNEW (Accessed March 3, 2014).

Office of Public Affairs. (2008, June 18). "Joint U.S.-China fact sheet: U.S.-China Ten Year Energy and Environment Cooperation Framework." U.S. Treasury Department. Retrieved from http://www.treasury.gov/initiatives/Documents/uschinased10yrfactsheet.pdf (Accessed March 3, 2014).

Oliva, M.J. (2008). "Climate change, technology transfer and intellectual property rights." International Centre for Trade and Sustainable Development. Retrieved from http://ictsd.org/downloads/2008/10/cph_trade_climate_tech_transfer_ipr.pdf (Accessed March 3, 2014).

Parkhe, A. (1993). "Strategic alliance structuring: A game theoretic and transaction cost examination of interfirm cooperation." *The Academy of Management Journal*, 36(4):794–829.

Patel, P., & Vega, M. (1999). "Patterns of internationalisation of corporate technology: Location vs. home country advantages." *Research Policy*, 28(2–3):145–55.

PCAST. (1997). *Report to the President on Federal Energy Research and Development for the Challenges of the Twenty-First Century*. President's Committee of Advisors on Science and Technology. Retrieved from http://clinton3.nara.gov/WH/EOP/OSTP/Energy/ (Accessed March 3, 2014).

PCAST. (1999). *Powerful Partnerships: The Federal Role in International Cooperation on Energy Innovation*. President's Committee of Advisors on Science and Technology. Retrieved from http://www.whitehouse.gov/sites/default/files/microsites/ostp/pcast-08021999.pdf (Accessed March 3, 2014).

PCAST. (2010). *Report to the President on Accelerating the Pace of Change in Energy Technologies through an Integrated Federal Energy Policy*. Executive Office of the President, President's Council of Advisors on Science and Technology Retrieved from http://www.whitehouse.gov/sites/default/files/microsites/ostp/pcast-energy-tech-report.pdf (Accessed March 3, 2014).

PEMEX. (2008). *Annual Report 2007/2008*. Petroleos Mexicanos. Retrieved from http://www.ri.pemex.com/index.cfm?action=content§ionID=135t&catID=12320 (Accessed March 3, 2014 Accessed March 3, 2014).

Peterson, P.F., Forsberg, C.W., Hu L-W., & Sridharan, K. (2013, June). "Integrated Research project FHR overview for DOE Nuclear Energy Advisory Committee." Presentation. Retrieved from http://energy.gov/sites/prod/files/2013/06/f1/FHRIRPPerPeterson_0.pdf (Accessed March 3, 2014).

PEW Charitable Trusts. (2011). "Who's winning the clean energy race?" PEW Charitable Trust, PEW Environment Group. Retrieved from http://www.pewenvironment.org/uploadedFiles/PEG/Publications/Report/G-20Report-LOWRes-FINAL.pdf (Accessed March 3, 2014).

Planning Commission. (2008). *Eleventh Five-Year Plan 2007–12*: Vol. I: *Inclusive Growth.* Retrieved from http://www.planningcommission.gov.in/plans/planrel/fiveyr/11th/11_v1/11th_vol1.pdf (Accessed March 3, 2014).

Pollack, H., & Congdon, M.B. (1974). "International cooperation in energy research and development." *Law and Policy in International Business*, 6(3):677–726.

Putnam, R.D. (1988). "Diplomacy and domestic politics: The logic of two-level games." *International Organization*, 42(3):427–60.

REN21. (2013a). *Renewables 2013: Global Status Report*. Paris: REN21 Secretariat.

REN21. (2013b). Renewables interactive map. Paris: REN21 Secretariat, Retrieved from http://map.ren21.net/ (Accessed March 3, 2014).

Reuters. (2008). "US says lax China IPR hampers clean tech trade." *Financial Express*, Beijing, China.

Ronstadt, R.C. (1978). "International R&D: The establishment and evolution of research and development abroad by seven U. S. multinationals." *Journal of International Business Studies*, 9(1):7–24.

Ruggie, J.G. (1975). "International responses to technology: Concepts and trends." *International Organization*, 29(3):557.

Sagar, A.D., & Anadon, L.D. (2009). "Climate change: IPR and technology transfer: Background paper prepared for the United Nations Framework Convention on Climate Change." *UNFCCC*, 1–48.

Sagar, A.D., & Gallagher, K.S. (2004). *Energy Research, Development, Demonstration and Deployment*. Cambridge, Mass.: Belfer Center for Science and International Affairs, John F. Kennedy School of Government, Harvard University. Retrieved from http://belfercenter.ksg.harvard.edu/files/energytechdd.pdf (Accessed March 3, 2014).

Sarewitz, D., & Pielke, R.A. (2007). "The neglected heart of science policy: Reconciling supply and demand for science." *Environmental Science & Policy*, 10(1):5–16.

Smits, R., & Kuhlmann, S. (2004). "The rise of systemic instruments in innovation policy." *International Journal of Foresight and Innovation Policy*, 1(1/2):4–32.

State. (2001). *Supplementary Handbook on the C-175 Process: Routine Science and Technology Agreements*. Office of Science and Technology Cooperation, Bureau of Oceans and International Environmental and Scientific Affairs, Department of State. Retrieved from http://www.state.gov/e/oes/rls/rpts/175/ (Accessed March 3, 2014).

State. (2010). "Financial resources and technology transfer." In *Fifth U.S. Climate Action Report*, pp. 99–120. Washington, D.C.: Department of State.

Stine, D.D. (2009). "Science, technology, and American diplomacy: Background and issues for Congress." Congressional Research Service. Retrieved from http://www.fas.org/sgp/crs/misc/RL34736.pdf (Accessed March 3, 2014).

Subcommittee on Global Change Research. United States Global Change Research Program. Retrieved from http://downloads.globalchange.gov/ocp/ocp2010/ocp2010.pdf (Accessed March 3, 2014).

Taylor, R. (2004). "NREL international program overview." Retrieved from http://www1.eere.energy.gov/solar/pdfs/nrel_international.pdf (Accessed March 3, 2014).

The Royal Society. (2011). "Knowledge, networks and nations: Global scientific collaboration in the 21st century." The Royal Society. Retrieved from http://royalsociety.org/uploadedFiles/Royal_Society_Content/Influencing_Policy/Reports/2011-03-28-Knowledge-networks-nations.pdf (Accessed March 3, 2014).

TPCC. (2010). "Renewable energy and energy efficiency export initiative." Trade Promotion Coordinating Committee, an interagency group chaired by the U.S. Secretary of Commerce. Retrieved from http://export.gov/build/groups/public/@eg_main/@reee/documents/webcontent/eg_main_024171.pdf (Accessed March 3, 2014).

Tyfield, D., Jin, J., & Rooker, T. (2010). "Game-changing China: Lessons from China about disruptive low carbon innovation." NESTA. Retrieved from http://www.nesta.org.uk/sites/default/files/game-changing_china.pdf (Accessed March 3, 2014).

UN COMTRADE. (2011). United Nations Commodity Trade Statistics Database. United Nations Statistics Division. Retrieved from http://comtrade.un.org (Accessed March 3, 2014).

UNDP. (2009). *Handbook on Planning, Monitoring and Evaluating for Development Results*. United Nations Development Programme. Retrieved from http://www.undg.org/docs/11653/UNDP-PME-Handbook-(2009).pdf (Accessed March 3, 2014).

UNFCCC. (2006). *Synthesis Report on Technology Needs Identified by Parties Not Included in Annex I to the Convention. United Nations Framework Convention on Climate Change*. Retrieved from http://unfccc.int/resource/docs/2006/sbsta/eng/inf01.pdf (Accessed March 3, 2014).

UNFCCC. (2008). *Bali Action Plan*. United Nations Framework Convention on Climate Change, 13th Conference of the Parties. Retrieved from http://unfccc.int/resource/docs/2007/cop13/eng/06a01.pdf#page=3 (Accessed March 3, 2014).

U.S. Congress. (2007). Energy Independence and Security Act of 2007. 110th Congress. Retrieved from http://www.gpo.gov/fdsys/pkg/BILLS-110hr6enr/pdf/BILLS-110hr6enr.pdf (Accessed March 3, 2014).

U.S. Congress. (2009). International Science and Technology Cooperation Act of 2009. 111th Congress, 1st Session. Retrieved from http://www.gpo.gov/fdsys/pkg/BILLS-111hr1736rfs/pdf/BILLS-111hr1736rfs.pdf (Accessed March 3, 2014).

USGCRP. (2009). *Our Changing Planet: The U.S. Climate Change Research Program for Fiscal Year 2010*. Washington, D.C.: United States Global Change Research Program. Retrieved from http://downloads.globalchange.gov/ocp/ocp2010/ocp2010.pdf (Accessed March 3, 2014).

Van de Ven, A., & Gordon, W. (1984). "The dynamics of interorganizational coordination." *Administrative Science Quarterly*, 29(4):598–621.

von Zedtwitz, M., & Gassmann, O. (2002). "Market versus technology drive in R&D internationalization: Four different patterns of managing research and development." *Research Policy*, 31(4):569–88.

World Bank. (2007). "Clean energy for development investment framework: Progress report of the World Bank Group Action Plan." Development Committee. Retrieved from https://openknowledge.worldbank.org/handle/10986/12559 (Accessed March 3, 2014).

World Nuclear Association. (2010). Nuclear power country briefings. Retrieved from http://www.world-nuclear.org (Accessed March 3, 2014).

WRI. (2010). SD-PAMs Database. Retrieved from http://projects.wri.org/sd-pams-database (Accessed March 3, 2014).

Young, O.R. (1989). *International Cooperation: Building Regimes for Natural Resources and the Environment*. Ithaca, NY: Cornell University Press.

Zhi, Q., Jun, S., Ru, P., & Anadon, L.D. (2013). "The evolution of China's national energy RD&D programs: The role of scientists in science and technology decision-making." *Energy Policy*, 61:1568–85.

6

Transforming U.S. Energy Innovation: How Do We Get There?

LAURA DIAZ ANADON, VENKATESH NARAYANAMURTI, AND MATTHEW BUNN

As we argued in Chapter 1, the United States and the world need a radical acceleration in the rate of energy technology innovation to meet the profound economic, environmental, and national security challenges that energy poses in the 21st century. If the U.S. government does not act now to improve the conditions for innovation in energy, even in times of budget stringency, it risks losing leadership in one of the key global industries of the future, and the world risks being unable to safely mitigate climate change and to reduce vulnerability to disruptions and conflicts – both domestic and international. Waiting is not an option.

The previous chapters of this book have laid out a broad agenda of policy changes that could help cause such a transformation in U.S. energy innovation. We believe that implementing these recommendations would bring the United States much closer to meeting the CASCADES criteria for an effective energy technology innovation strategy that we laid out in Chapter 1 – an approach that is comprehensive, adaptable, sustainable, agile, diversified, equitable, and strategic. In this chapter, we summarize the recommendations of the previous chapters, build on crosscutting themes that run through them all, and offer a preliminary discussion of the steps that would be needed to put the United States on the trajectory we suggest.

6.1. A Major Expansion in Energy Technology Innovation Investments, with More Stability and Better Targeting

As discussed in Chapter 2, the U.S. government should improve its approach to making decisions about energy RD&D investments in different technology areas. We have developed an approach to support these decisions and concluded that dramatically expanding the Department of Energy (DOE)'s investment in energy research, development, and demonstration (RD&D), focused on a broad portfolio of different energy technologies and stages of innovation, is justified by the expected social benefits.

In particular our results support an energy RD&D budget in the range of $10-$15 billion in the seven technology areas considered.[1] Experts in the energy technologies covered by the analysis almost unanimously recommended increasing energy RD&D in their areas by two to three times, to seize the technological opportunities that now exist. Our economic modeling suggests that an investment of a few extra billion per year could develop technologies that could save the economy hundreds of billions of dollars per year by 2050 in scenarios where there are stringent policies limiting how much carbon can be emitted. Our analysis shows that investments should be targeted on a broad portfolio of technologies, to maximize the chances of achieving major breakthroughs.

Our method combines expert assessments of the improvements in cost and performance of particular technologies that could be achieved through RD&D investments with economic modeling to estimate what benefits such improvements could bring for the overall U.S. energy economy – whether judged by reduced cost, increased ability to reduce carbon emissions, or increased energy security. Crucially, our method also takes at least a first step toward incorporating the enormous uncertainties that exist in estimating what the results of RD&D will really be – though the potential for breakthroughs beyond what experts currently imagine remains difficult to incorporate in any model. The modeling results presented in Chapter 2 showing decreasing marginal returns to RD&D in terms of benefits to society reflect the fact that most of the experts surveyed expected decreasing marginal returns in terms of technology improvements with increasing RD&D investments. This suggests that our methodology could be used to estimate the level of RD&D investment that would maximize social benefits. It is important to note that any optimal level would have to be regularly reassessed as technologies evolve. Indeed, the specific figures are less important than the broad conclusion that a major increase in spending is needed.

To be effective, these investments must be sustained over a period of years, in a stable and predictable approach (while allowing flexibility to adapt to changing circumstances). In the past, DOE's energy research, development, and demonstration (RD&D) investments overall, and in different technology areas in particular, have been very volatile (as shown in Figures 2.2 and 2.3). Rapid shifts in funding priorities – which extend to some forms of support for the deployment of energy technologies[2] –

[1] As discussed in Chapter 2, in 2011, we recommended an increase of U.S. government energy RD&D funding for these seven technology areas from $2.1 billion to $5.2 billion. We recommended a total of $10 billion including funding for energy RD&D areas beyond the seven covered in this report, ARPA-E, and Basic Energy Sciences. Further modeling for this book and other work (Chan and Anadon, 2014) indicates that even higher levels of funding, amounting to $10 to $15 billion for the seven areas we surveyed would have positive overall benefits for society. This would imply a total of $15 to $20 billion for all energy RD&D programs (not just the seven we focused on for the modeling analysis), ARPA-E, increase in funding than we called for in 2011, though of course this increase would be phased in over a number of years.

[2] Although the analysis of federal deployment policies, such as production tax credits (PTC) and investment tax credits for wind and solar power and Renewable Fuels Standards, was not the focus of this work, these policies have also experienced volatility and uncertainty. Previous research focused on the wind PTC has shown that the

undermine innovation. Key reasons for these shifts in include changing perceptions of the attractions of different technologies (both in the executive branch and in Congress) and political disagreements on the role of government in technology development ("the private sector and free markets will decide"). By whatever means proves to be practicable, the U.S. government should seek to adopt a more stable approach to making these long-term investments.

The method we present may be more important than the specific funding allocations we recommend, which are sure to change as time goes on. We believe decision makers at DOE headquarters and in its program offices should conduct integrated analyses of the potential benefits of different RD&D investments under different scenarios to help guide their investment decisions.[3] As highlighted by a report from a committee of the National Research Council (NRC, 2007), prospective benefits analysis can help to craft more efficient RD&D investment policies. Developing transparent and consistent assessments of the expected benefits of various investment portfolios, including accounting for the relevant uncertainties, can inform the RD&D decision-making process by highlighting dependencies, uncertainties, and benefits that may not be obvious. A credible and independent way of understanding the impact of an investment portfolio may also help the persuasiveness of energy RD&D proposals to other members of the executive branch and to Congress. We have developed and implemented a tool for such benefits analysis, relying on expert elicitations, energy-economic models, and optimization algorithms. For DOE to use this method successfully would require support at the highest levels of the department and a clearly independent group to conduct credible expert elicitations and modeling.

Governments do not have unlimited resources and must prioritize. But despite deficits that have dominated the discourse over the past few years, the United States cannot afford to forego the long-term investments that will improve its competitiveness in this multi-trillion-dollar market and its national security, while reducing both greenhouse gas emissions and other environmental hazards. Expanded investments in energy RD&D are likely to be essential if we are to avoid climate catastrophes at reasonable cost.

If it is too difficult to provide expanded, stable investments through the normal budget process, there are a variety of other options for funding energy RD&D, some of which have been implemented successfully on a smaller scale in the past. For example, the Gas Research Institute (GRI) discussed in Chapter 1, which funded RD&D managed by a board that included both industry and government representatives, was

year-to-year extension of such policies creates uncertainties that deter private investment (Barradale, 2010). The Renewable Fuel Standard2 (RFS2) provides an assurance of a market under most ordinary circumstances. However, as detailed in a 2011 NRC report (NRC, 2011): "language in the legislation gives EPA the right to waive or defer enforcement of RFS2 under a variety of circumstances," which contributes to policy uncertainty.

[3] Note that we are *not* arguing that the use of this integrated analysis should serve as a support tool to make investment decisions in all DOE energy RD&D–for example, investments in Basic Energy Sciences and ARPA-E are not covered.

established with a small charge on interstate transportation of natural gas, approved by the Federal Energy Regulatory Commission (FERC) and supported by industry. Similar initiatives could be launched today, such as a FERC-approved surcharge on electricity, intended to fund development of better approaches to generating electricity for the future (PCAST, 2010). Such off-budget opportunities should be considered and pursued as complements to funding provided through the regular appropriations process.

6.2. Matching Energy Technology Push to Market Pull

The U.S. federal government should implement policies that create market incentives to develop and deploy new energy technologies, including policies that have the effect of creating a substantial price on carbon emissions and sector-specific policies to overcome other market failures.

The modeling presented in Chapter 2 makes clear that both expanded RD&D investments and a substantial price on carbon emissions are necessary if the United States is to meet the climate change challenge and reduce dependence on imported oil at reasonable cost. Increased energy innovation investment alone, with no policies in place to accelerate the adoption of the resulting technologies and further spur private sector innovation, is very unlikely to reduce U.S. greenhouse gas emissions by more than a few percent by 2050 – even with huge increases in RD&D and optimistic technology assumptions. Similarly, carbon caps or prices alone, with no increase in energy RD&D, would make the resulting energy system more costly. A clean energy standard could contribute toward the goal of reducing carbon emissions, but by concentrating the effort on only a portion of the sources of emissions, it would be less effective than a broader carbon policy. Together, well-integrated "technology push" from increased RD&D and "demand pull" from a carbon price and sector-specific policies (such as building codes and efficiency standards) have the potential to accomplish what neither can do alone.

At the same time, it is important to recognize that crafting demand-side programs is not an easy task, either politically or technically. Historically, the United States has relied on a variety of tools to incentivize deployment of particular energy technologies. It has used mainly investment and production tax credits for renewable power technologies, for oil and gas exploration, and for advanced coal power; federal insurance and a variety of supports for nuclear power; mandates for biofuels; and standards for efficiency. Often the particular measures adopted had more to do with lobbying by special interests than with an optimal approach to promoting a broad portfolio of technologies – and they often have been quite disconnected from the technology push of RD&D investments, rather than being designed to create an integrated path from invention to deployment. And these incentives have been attacked for a lack of clarity in their goals and effectiveness, and in the case of the tax credits for renewables for their short-term nature. Most importantly, they are not part of a deliberate energy policy that mobilizes different agencies and policy tools to achieve specific goals. It becomes paramount to create a comprehensive federal energy policy, as others have recommended (PCAST, 2010).

Of course, the federal government is not the only public actor whose policies matter in encouraging deployment of new energy technologies: states and cities play important roles as well. In addition to making their own RD&D investments (which by some estimates were around $3 billion in 2008) many states have their own renewable portfolio standards in place, which help new technologies move down the learning curve by requiring utilities to deploy a certain portion of renewable energy even if the technologies are not yet cost-competitive with fossil fuels. For businesses operating across large geographic areas, however, a single national requirement would be more efficient than having to navigate different requirements and definitions in each state. Efficiency codes and other measures to encourage deployment are often established at the municipal level. Overall, cities and states provide an invaluable opportunity for policy experimentation (NRC, 2010), as demonstrated by the regional cap-and-trade programs. But only the federal government has the ability to shape energy prices and energy technology markets on a national scale, increasing the overall efficiency and impact of the effort.

6.3. Reforming Energy Innovation Institutions

As discussed in Chapter 3, to maximize the return on its investment in energy RD&D, the U.S. government must ensure that its energy innovation institutions are as efficient and effective as they can be. Constantly shifting funding, DOE headquarters' micro-management, diffuse missions, risk-averse cultures, new contracting approaches that have diluted the focus on key national objectives, an artificial separation between basic and applied research, and insufficient ability to connect to the private sector have undermined the efficiency of U.S. energy RD&D institutions.

To resolve these issues, the U.S. government should strengthen its energy innovation institutions, particularly the national laboratories, by giving them clear missions and direction informed by, and linked to, a larger systems perspective, with a focus on a public good; leadership with proven technical and managerial excellence and a vision of the role of the institution in the overall innovation ecosystem; considerable management authority and flexibility with clear accountability for results; stable and predictable funding that allows for a thorough exploration of technical opportunities; flexibility to deploy funds quickly into promising areas when opportunities arise an entrepreneurial culture willing to invest in high-risk, high-payoff projects; and opportunities to lend their insights to the design of the policies and approaches they are helping to implement, including public–private partnerships (SEAB, 1995; Narayanamurti et al., 2009; PCAST, 2010; DOE, 2011). We recommend a range of reforms, including an increase in lab-directed funds to allow lab directors to provide seed funds for promising areas, incentives for entrepreneurialism and risk-taking for lab scientists and technicians, and a revision of contracting in the labs.

Advanced Research Project Agency-Energy (ARPA-E) represents a major step in the right direction, emphasizing technical excellence, rapid adaptation, and high-risk, potentially high-payoff projects. Policymakers should provide consistent, multiyear support for ARPA-E and should not be deterred by a few failures – if there were no failures, the effort would not be pursuing the high-risk projects that have the highest potential payoffs. The Energy Innovation Hubs and the Energy Frontier Research Centers are also promising.

As highlighted in Chapter 3, in some cases there is a compelling need for public–private partnerships to support commercial-scale demonstration of a new technology, when it is at a stage where it is almost ready for commercial deployment but the risk is still too high for the project to be financed in private markets. Although DOE has attempted such demonstrations in the past – and is continuing to work on the major FutureGen demonstration of carbon capture and sequestration technology – it does not currently have institutions that are well-structured for this type of work, leaving an important gap in the U.S. system for supporting energy innovation, despite new institutions such as ARPA-E and the Hubs (Narayanamurti et al., 2011). Such demonstrations are likely to be expensive – potentially in the range of billions of dollars for a single project – and the government and the private sector are both likely to have to contribute. Innovative public–private arrangements may be needed to implement such demonstrations effectively. Moreover, the overall funding recommendations just described may not be sufficient to support the full range of large-scale demonstrations that may be needed; funding for an institution intended to carry out such demonstrations might have to be provided in addition to the funding recommended above.

One particularly needed reform is to remove the stove piping that is common between different technology programs at DOE and between basic and applied research. As one of the authors has argued elsewhere, even using the language of "basic" and "applied" research obscures the important feedbacks going both ways between them and has a negative impact on how the RD&D enterprise is designed and shaped (Narayanamurti et al., 2013). This division affects the type and timeframe of the research that is happening on the ground, the type of knowledge that gets shared, and the type of interactions that take place.[4] This distinction should be replaced with an integrated approach – similar to that which once existed at Bell Labs, which took technologies such as transistors and lasers from fundamental research at the frontiers of science to commercial products that transformed industries.

[4] Even though in some programs there are formal linkages between, for example, the storage-related research in the Basic Energy Sciences (BES) funded program and the vehicles program in EERE, and in many cases informal linkages between researchers and program managers, the arbitrary division between basic and applied research is not completely bridged.

Just as the basic–applied divide should be viewed as an interdependent ecosystem with bidirectionality, the boundaries between the labs and the private sector should be more porous. National Renewable Energy Laboratory (NREL), for example, does not provide its researchers with opportunities to spend one to two years working in a different environment to refresh their knowledge and gain a better understanding of the marketplace. Sandia National Laboratory's program allowing such exchanges may serve as a model to other labs.

The government-owned, contractor operated (GOCO) system of the labs was put in place in the 1940s and 1950s to insulate the labs from political pressures and allow them to run efficiently. These labs were often run by leading research universities or private firms with strong technical credentials such as AT&T and DuPont, using "no profit, no fee" contracts. Today, major companies expecting to make significant profits from running these facilities manage many of the labs. While these labs retain a major focus on service to the nation, their management now inevitably also focuses on what is good for the firm, and the sense of mission has been somewhat diluted – a problem exacerbated by putting the contracts up for bidding every five years, leading to a major focus on winning the next bid rather than accomplishing the long-term mission. In addition, companies running labs for a profit inevitably have a major emphasis on cost control among their managers, which may dilute the focus on technical excellence that may be present in universities or other leading research organizations. The shift to for-profit management has also contributed to micromanagement from DOE headquarters, where officials often see the labs as "just another contractor," deserving little role in policy development or advice. At the same time, since the 1990s, the contracts to manage the labs have often been written with strong incentives for focusing on operational rules at the expense of the primary technical mission. The effort to insulate the labs from political pressures has also weakened owing to the politicization of the appropriations process. This is dramatically illustrated by the large year-to-year fluctuations in the budget of NREL. At NREL, the level of long-term lab-directed research and development (LDRD) funding is also well below the allowable costs; this hinders NREL's ability to attract high-quality people interested in long-term R&D.

More broadly, the current funding structure of the "more applied" labs such as NREL does not allow the director or project managers to have sufficient funds and discretion to react quickly to new research opportunities. Working with firms and other institutions is often a cumbersome process, requiring a flexible ability to shift funds when needed. By contrast, an environment of budget instability inevitably focuses managers on trying to secure more funding. In the current environment, the way to do this is often by promising goals that can be accomplished with high confidence, which are by nature incremental rather than transformational. This is partly a function of the fact that many policymakers do not understand the nature of research (in fact the best research is that which stretches the boundary of what was thought possible), but the fact of the matter is that the timescales of the research at the labs, as well as the interaction with industry, the ability to react to new opportunities, and the technical excellence of the management, all need to change.

With initiatives such as ARPA-E and the Energy Innovation Hubs, the U.S. energy innovation system is being reshaped. This brings not only a rare opportunity, but also a responsibility, to improve this system's efficiency and effectiveness – and the principles we offer in this book can help make that happen.

6.4. A Strategic, Partnership-Based Approach to the Private Sector

As outlined in detail in Chapter 4, the U.S. government should take a strategic approach to working with the private sector on energy innovation, expanding incentives for the private sector to invest in its own energy innovation and focusing on the particular strategies likely to work best in each case.

The private sector has a critical role to play, as private firms control the vast majority of the energy infrastructure in the United States and perform the majority of energy R&D. Indeed, our research shows that private sector innovation in energy is more widespread than previously understood. Yet, today, DOE does not have an overall strategy for its interactions with the private sector. It does not collect and analyze the data necessary to learn from the experience of past projects about what mechanisms for collaboration work best and under which circumstances across the whole Department. We propose a new approach in which that data would be collected and analyzed, enabling learning-by-doing. Our research also shows that the demand pull policies we recommend will enhance incentives for private sector investment in energy innovation, which should also be strengthened through other policies, ranging from support for large-scale technology demonstrations to investment in training the next generation of energy technologists.

Recent NSF data show that in 2010, more than 11,000 U.S. firms invested a total of at least $16 billion in energy-related R&D. The firms doing R&D related to energy production and use represent almost a quarter (23%) of all the firms doing R&D of any kind in the United States. The fact that such a large percentage of firms active in R&D are engaged in energy-related R&D is a testament to the ubiquity of energy and energy concerns throughout the economy. Our pilot survey further showed that an even greater number of business establishments (about 260,000) are active in energy R&D and that two out of three establishments were expecting to recoup their energy RD&D investments in just two to three years. The widespread nature of private energy RD&D also highlights the enormous potential impact of additional nudges to invest in energy innovation, which could lead to substantial additional private investment – and resulting faster evolution of energy technologies.

While the Internet revolution and the rise of computers for everyone were sparked by early government investments, from Defense Advanced Research Projects Agency

(DARPA) and other parts of the defense establishment, they also relied crucially on an open market where there were large payoffs for developing new products, and the new inventions did not have to compete with incumbents that could provide similar service and had already had decades in which to optimize their performance, reduce costs, and establish infrastructure and institutional support. In energy, all of those barriers to change are very real. Even if U.S. government energy RD&D investments increased to at least $10 billion a year, as we recommend, and even if the government's energy innovation institutions were reformed in the ways we suggest, there would still be a crucial need for a long-term strategy to attract industry players to invest in both RD&D and in commercialization. As noted previously, this long-term strategy needs a demand-side component that creates an expectation that companies can make substantial profits if they provide cheaper, more effective ways to address U.S. energy challenges. Indeed, our survey of energy innovation in U.S. businesses singled out energy prices as the most important factor shaping their energy R&D investments, followed by federal grants. The research presented in Chapter 4 on DOE grants and cooperative agreements with private firms and universities reveals that no coherent strategy for working with the private sector is yet in place.

Finally, Chapter 4 highlights the need to strengthen the interaction between the government and the private sector throughout the energy innovation system. Although members of the private sector testify to Congress, participate in advisory boards, and respond to requests for information, it would be useful to integrate private sector perspectives more fully in the design of policies and research programs. Hence, as discussed in Chapter 4, we propose that DOE establish an Energy Innovation Advisory Board (EIAB) that would include representatives from the private sector, academia, the national laboratories, and other key stakeholders, to advise on how best to accelerate private sector energy technology innovation. The board's mission should include not only advice on DOE's investments in energy technologies, but also on the broad spectrum of U.S. government policies that affect the pace of energy technology innovation.

6.5. Strengthening International Cooperation on Energy Technology Innovation

As discussed in Chapter 5, the U.S. government should undertake a strategic approach to energy RD&D cooperation with other countries, to leverage the knowledge, resources, and opportunities available around the world, incorporating both top-down strategic priorities and investment in new ideas arising from the bottom up.

The energy challenges of the 21st century are global, not limited to the United States. World energy markets and energy technology development are rapidly changing as globalization progresses and new players emerge. Today, total government-controlled energy RD&D in just six emerging or transition countries (Brazil, Russia, India, China, Mexico, and South Africa), counting investments by state-owned enterprises, is as large or larger than government-sponsored energy RD&D in all of the developed countries combined. These new realities create both new competitive challenges and new opportunities, requiring the United States to ensure that high-value cooperation opportunities are not missed because of the sometimes ad hoc and distributed nature of existing cooperation activities.

As discussed in Chapter 5, we recommend a strategic approach to balancing international competition and cooperation in energy RD&D, focused on answering key questions related to the benefits and risks of proposed projects – while keeping in mind that the process of cooperation itself can provide opportunities for discussions and joint work that help address some concerns over unfair competition. Recognizing that innovators on the ground may be most aware of new opportunities as they arise, we recommend that each of the departments and agencies financing energy RD&D set aside a small portion of its funds to support international energy RD&D projects suggested from the bottom up. In parallel, we recommend that a new top-down interagency task force be established to identify and mobilize resources behind key priorities, provide overall direction to the effort, minimize overlaps, and seize synergies among the efforts underway. Finally, we recommend that a new approach be established to collecting and analyzing the data needed to support an overall strategy and to learn what approaches to international energy RD&D cooperation work best under what circumstances.

6.6. Cross-Cutting Themes

Throughout this book, we have argued that, to meet the large energy challenges facing the United States and the world, the U.S. government must do more and better to promote innovation across a range of energy technologies. As we explored the issues surrounding the best size and scale of U.S. government energy RD&D investments, the institutions intended to promote energy innovation, the steps needed to strengthen private sector innovation, and the role of international cooperation in energy RD&D, several cross-cutting themes emerged.

6.6.1. The Need for a Federal Energy Strategy

The first crosscutting theme is the urgent need for the U.S. government to establish an overall energy strategy, and a long-term strategic approach to its energy technology innovation efforts in support of that strategy (e.g., PCAST, 2010). Such a strategy is essential for providing a unifying sense of mission, a rationale for investments and other policies, and a level of stability in the government's approach that is currently lacking.

The wide range of public and private actors involved in energy technology innovation further emphasizes the need for a guiding strategy that enunciates clear goals.

This would help the huge number of federal departments and agencies, state and municipal governments, firms, laboratories, universities, and partners in other countries that are involved in U.S. energy innovation effort to work together more efficiently toward common goals.

The *Quadrennial Technology Review* that DOE undertook in 2011 was a first step, identifying multiyear programs for different technologies on the basis of technical assessment of their potential. But a strategy for the federal government as a whole is necessary. This federal strategy should navigate the difficult tension between stability and the flexibility (or adaptability) needed to use new information. A federal strategy would also help with (and require) an improved coordination among federal agencies and with states. Again, the *Quadrennial Technology Review* helps provide coordination within DOE, but ultimately a government-wide review (such as the *Quadrennial Energy Review* recommended by the President's Committee of Advisors in Science and Technology [PCAST] and initiated by the President in January 2014) will be needed to align DOE programs with government-wide initiatives. For example, a price on carbon or a clean-energy standard will affect private sector activity in energy innovation and the investments that DOE should make.

The need for a more stable, strategic approach comes out in every chapter of this book – when analyzing the volatility of DOE's investments in different technology programs, the lack of an integrated approach in DOE's process for making recommendations about these investments, the rapid shifts and lack of overall strategy in DOE's approach to funding and partnering with private firms, and the myriad and shifting approaches underway for international cooperation in energy RD&D. At the same time, private firms need long-term price signals and reduced policy uncertainty if they are to be able to effectively accelerate innovation in energy technologies. A federal strategy would contribute to the creation of stable long-term policies that allow industry to plan and make investments and that provide researchers with time to collect sufficient information and to explore different research avenues efficiently.

6.6.2. Learning from Experience

The second crosscutting theme that comes out throughout this book is the need for an energy technology innovation strategy that is able to learn from experience, improving its effectiveness over time. In the military sphere, it is often said that no battle plan ever survives contact with the enemy. In much the same way, plans for energy technology innovation need to learn from and adapt to the realities they confront as they proceed – while maintaining a level of stability necessary for planning investments and policies. The complexity, heterogeneity, and ubiquity of energy technologies and markets imply the need for some humility and the ability to learn and adapt the strategy over time and better tailor different tactics to different technologies, actors, and markets.

Today, DOE program managers are often unaware of which approaches worked and which did not to develop different types of technologies with different, making it difficult to strengthen these efforts by learning lessons from past experience in areas outside of their own. This theme recurs whether the topic is allocating RD&D investments, managing the national laboratories and other energy innovation institutions, managing cooperation with the private sector, or optimizing the U.S. approach to international cooperation on energy technology innovation.

For private sector interactions in particular, current institutional arrangements provide little guidance or evidence across the whole DOE to help managers make decisions about what type of approach – for example, grants, cooperative agreements, R&D consortia, work for others, user facilities, and Cooperative Research and Development Agreements (CRADAs) – to use in each circumstance. Although program monitoring and evaluation efforts of the Office of Energy Efficiency and Renewable Energy are beginning to address this knowledge gap, a DOE-wide effort is needed. Interactions with private firms are often more difficult and bureaucratic than they should be and, at the same time, in some cases the interaction is not strong enough – partly owing to the (often correct) perception that some firms have that the federal government is not a stable partner, which goes back to the need for a strategy and stability.

More broadly, in Chapters 1 to 5, we outline specific institutional approaches to collecting and analyzing the information needed to learn by doing for each of the categories of U.S. government energy innovation effort addressed in this volume.

6.6.3. Creating the Workforce for the Future

A third theme is that none of this can be accomplished without the right people – people with an excellent understanding of the technical issues combined with managerial talent, an ability to work effectively with others, and a vision of where the effort should go. One key role for the federal government in energy innovation is to support training the next generation of world-class scientists and engineers to meet the world's energy challenges.

To this end, we recommend that DOE identify training the future energy technology workforce in the United States as an explicit mission, and fund programs designed to accomplish that goal. This is an area that DOE is particularly well suited to contribute to, given that relatively small resources could make a big difference. There are many options that are already contributing to workforce development, including creating more Energy Frontier Research Centers (EFRCs) and expanding the Hubs to increase the component devoted to engaging undergraduate and graduate students as well as postdoctoral fellows. DOE could also create additional long-term funding for multidisciplinary programs at universities. In the nuclear sphere in particular, Congress and the administration have shortsightedly cut funding targeted on university programs; this funding should be restored and expanded, and similar programs established to train experts in other areas of energy technology.

The increased interaction with the private sector we recommend will be essential to this workforce development effort, in order to identify areas of near-term need. But DOE should also keep a distinct long-term view to provide scientists and engineers for the technologies of tomorrow, as well as a distinct focus on developing interdisciplinary scholars. Although universities are beginning to recognize the importance of collaboration across disciplines, the interests of individual departments as well as the criteria for academic success often clash with efforts to conduct interdisciplinary scholarship (ARISE2, 2013).

6.7. The Path Ahead

Together, the recommendations in this book can help the United States government meet the CASCADES criteria outlined in Chapter 1 for an effective energy innovation approach. To transform U.S. energy technology innovation that is required will require an approach that is:

- *Comprehensive*, covering the full suite of technology stages and actors in the energy innovation ecosystem
- *Adaptive*, able to learn and shift course in response to experience and changing conditions
- *Sustainable*, maintaining the stability needed to make progress over time toward long-term objectives
- *Cost-effective*, focusing on the strategies designed to offer the most progress per dollar invested
- *Agile*, able to reallocate resources to respond to new opportunities (and failures) as they arise
- *Diversified*, covering all technological pathways that offer substantial promise of contributing to U.S. energy goals
- *Equitable*, taking an objective, fair approach to different technologies and actors in the system, rather than carving out special favors for particular interests
- *Strategic*, targeted effectively on a clear set of goals designed to address the most critical energy challenges the United States faces

In this volume, we have presented new data and analysis on various aspects of the U.S. energy innovation system with a focus on DOE. We have also laid out an agenda for energy innovation policy that can meet these criteria and transform the pace at which the United States develops and deploys the energy technologies of the future.

Implementing this agenda will require sustained high-level leadership from the administration and the Congress. In particular, to be sustainable, it will require bipartisan support and cooperation – a commodity that has been in short supply in recent years.

But there are some signs that building a bipartisan coalition of support for such an agenda may be possible. Both the Bush administration and the Obama administration proposed substantial increases in funding for energy RD&D, and both supported reforming U.S. energy innovation institutions, from creating ARPA-E to initiating major loan guarantees for certain low-carbon energy projects. The legislation to appropriate funding for these new institutions as received bipartisan support – and indeed, the subcommittees handling funding for energy RD&D, are among the most bipartisan groupings in the U.S. Congress. There are strong supporters of increased investments in energy RD&D – and even some supporters of effective demand pull policies – on both sides of the U.S. political debate. Indeed, until the mid-2000s, there was reasonably strong bipartisan support for market-based approaches such as a carbon tax or a carbon cap and trade system – and as the impact of climate change continues to grow, the politics of such measures may change again. Although some technologies find greater resonance with one U.S. political party than another, the fact is that both parties include strong supporters of many of the key technologies discussed in this volume – from nuclear to solar to wind to biofuels to efficiency to storage.

The need for the United States to develop a broad panoply of new and improved energy technologies will only become clearer over time, as other countries continue to invest, climate impacts continue to worsen, the impact of fine particulates and other damages from energy production and use becomes more stark, and the fact that cheap oil and gas will not last forever becomes more clear. U.S. policymakers should be able to develop a consensus that they do not want the United States to be left behind by China and other countries in the energy technology markets of the future, that they want technologies that would strengthen U.S. energy security – and, ultimately, that it is worth investing in the insurance represented by developing lower-cost options for reducing the world's carbon emissions in the future.

In short, there is reason to hope that over the medium term, an agenda such as the one laid out in this volume can find sustainable bipartisan support. It is our hope that by laying out the case for an expanded energy innovation agenda and suggesting approaches that can help make it most effective, this book can contribute, in a small way, to bringing such support about. The energy challenges the United States and the world now face are staggering, and a transformed approach to energy innovation will be an essential part of the effort to meet them.

References

Anadon, L.D. (2012). "Missions-oriented RD&D institutions in energy between 2000 and 2010: A comparative analysis of China, the United Kingdom, and the United States." *Research Policy*, 41(10):1742–56.

ARISE2. (2013). *ARISE2: Advancing Research In Science and Engineering2*. American Academy of Arts and Sciences. Cambridge, Mass. Retrieved from http://www.amacad.org/arise2.pdf (Accessed February 12, 2014).

Barradale, M.J. (2010). "Impact of public policy uncertainty on renewable energy investment: Wind power and the production tax credit." *Energy Policy*, 38:7698–709.

BESAC. (2007). "Directing matter and energy: Five challenges for science and the imagination." Washington, D.C.: U.S. Department of Energy. Retrieved from www.osti.gove/bridge/servlets/purl/935427-9dklDm/935427.pdf (Accessed June 5, 2012).

Bonvillian, W.B., & Van Atta, R. (2011). "ARPA-E and DARPA: Applying the DARPA model to energy innovation." *Journal of Technology Transfer*, 36:469–513.

CCNEWS. (2011). "15 state-owned enterprises laid the foundation stone of the future science and technology base at Changping District, Beijing." CCNEWS. Retrieved from http://ccnews.people.com.cn/GB/ 14296353.html (Accessed April 20, 2011).

Chan, G., & Anadon, L.D. (2014). "A new method to support public decision-making on R&D: An energy example." Submitted.

Choi, H., & Anadon, L.D. (2014). "The role of the complementary sector and its relationship with network formation and government policies in emerging sectors: The case of solar photovoltaics between 2001 and 2009." *Technology Forecasting and Social Change*, 82:80–94.

Chu, S. (2010). "China's clean energy successes represent a new "Sputnik moment" for America." Washington, D.C.: U.S. Department of Energy (Speech at the National Press Club on November 29, 2010).

DOE. (2011). *Quadrennial Technology Review*. U.S. Department of Energy, Washington, D.C. Retrieved from http://energy.gov/sites/prod/files/QTR_report.pdf (Accessed February 12, 2014).

DOE. (2013). "The history of solar PV." Washington, D.C.: U.S. Department of Energy, Office of Energy Efficiency and Renewable Energy. Retrieved from http://www1.eere.energy.gov/solar/pdfs/solar_timeline.pdf (Accessed February 12, 2014).

Duderstadt, J., Muro, M., Was, G., Sarzynski, A., McGrath, R., Corradini, M., Katehi, L., & Shangraw, R. (2009). *Energy Discovery-Innovation Institutes: A Step Toward America's Energy Sustainability*. Washington, D.C.: Metropolitan Policy Program, Brookings Institution.

EIA. (2013). *Annual Energy Outlook*, AEO2013 Early Release Overview. Energy Information Administration. Washington, D.C.: U.S. Department of Energy.

Gruebler, A. (2012). "Grand designs: Historical patterns and future scenarios of energy technological change. Historical case studies of energy technology innovation. Case study in chapter on Policies for the Energy Technology Innovation System (ETIS). In *Global Energy Assessment: Toward a Sustainable Future*, pp. 39–53, edited by A. Gruebler, F. Aguayo, K.S. Gallagher, M. Hekkert, K. Jiang, L. Mytelka, L. Neij, G. Nemet, & C. Wilson. Cambridge: Cambridge University Press.

Hoppmann, J., Peters, M., Schneider, M., & Hoffmann, V.H. (2013). "The two faces of market support – how deployment policies affect technological exploration and exploitation in the solar photovoltaic industry." *Research Policy*, 42(4):989–1003.

Jiabao, W. (2011). *Report on the Work of the Government*. Beijing, China: Xinhua News Agency.

Kempener, R., Anadon, L.D., & Condor, J. (2010). "Governmental energy innovation investments, policies, and institutions in the major emerging economies: Brazil, Russia, India, Mexico, China, and South Africa." Cambridge, Mass.: Energy Technology Innovation Policy Group, Belfer Center for Science and International Affairs, John F. Kennedy School of Government, Harvard University.

Li Zheng, J. (2011). "Coal in China." Presentation at the 10th Annual Meeting of the Carbon Mitigation Initiative on Energy and Carbon Intensity in China, April 7, 2011. Princeton, N.J.: Princeton University.

MIT. (2011). "Appendix 8A: Natural gas RD&D background." In *The Future of Natural Gas*. Cambridge, Mass.: Report from the Energy Initiative, Massachusetts Institute of Technology. Retrieved from http://mitei.mit.edu/system/files/NaturalGas_Appendix8A.pdf (Accessed February 12, 2014).

Narayanamurti, V., Anadon, L.D., & Sagar, A.D. (2009). "Institutions for energy innovation." *Issues in Science and Technology*, Fall:57–64.

Narayanamurti, V., Anadon, L.D., Breetz, H., Bunn, M., Lee, H., & Mielke, E. (2011). *Transforming the Energy Economy: Options for Accelerating the Commercialization of Advanced Energy Technologies*. Report. Cambridge, Mass.: Energy Technology Innovation Policy research group, Belfer Center for Science and International Affairs, Harvard Kennedy School. February. Retrieved from http://belfercenter.ksg.harvard.edu/publication/20729/ (Accessed February 12, 2014).

Narayanamurti, V., Odumosu, T., & Vinsel, L. (2013). "RIP: The basic applied divide." *Issues in Science and Technology*, XXIX (2, Winter).

NAS. (2005). *Rising Above the Gathering Storm: Energizing and Employing America for a Brighter Economic Future*. Washington, D.C.: National Academies of Sciences, National Academies Press.

Nemet, G.F. (2009). "Interim monitoring of cost dynamics for publicly supported energy technologies." *Energy Policy*, 37(3):825–35.

NRC. (2007). *Prospective Evaluation of Applied Energy Research and Development at DOE (Phase Two)*. National Research Council. Washington, D.C. National Academies of Sciences, National Academies Press.

NRC. (2010). *Limiting the Magnitude of Future Climate Change*, p. 2. Washington D.C.: National Academies Press.

NRC. (2011). *Renewable Fuel Standard: Potential Economic and Environmental Effects of U.S. Biofuel Policy*, p. 271. National Research Council. Washington D.C.: National Academies Press.

PCAST. (2010). *Report to the President on Accelerating the Pace of Change in Energy Technologies Through an Integrated Federal Energy Policy*. Washington, D.C.: President's Council of Advisors on Science and Technology. Executive Office of the President.

SEAB. (1995, February). *Alternative Futures for the Department of Energy National Laboratories*. Also known as "The Galvin Report." Secretary of Energy Advisory Board. February. Retrieved from http://www.lbl.gov/LBL-PID/Galvin-Report/Galvin-Report.html (Accessed February 12, 2014).

Zhi, Q., Jun, S., Peng, R., & Anadon, L.D. (2013, October). "The evolution of China's National Energy RD&D programs: The role of scientists in science and technology decision making." *Energy Policy*, 61:1568–85.

Index